BURLEIGH DODDS SCIENCE: INSTANT INSIGHTS

NUMBER 109

Alternative sources of protein for poultry

I0130671

burleigh dodds
SCIENCE PUBLISHING

Published by Burleigh Dodds Science Publishing Limited
82 High Street, Sawston, Cambridge CB22 3HJ, UK
www.bdspublishing.com

Burleigh Dodds Science Publishing, 1518 Walnut Street, Suite 900, Philadelphia, PA 19102-3406, USA

First published 2025 by Burleigh Dodds Science Publishing Limited
© Burleigh Dodds Science Publishing, 2025. All rights reserved.

British Library Cataloguing in Publication Data
A catalogue record for this book is available from the British Library

ISBN 978-1-83545-010-9 (Print)
ISBN 978-1-83545-011-6 (ePub)

DOI: 10.19103/9781835450116

Typeset by Deanta Global Publishing Services, Dublin, Ireland

Contents

Series list

Acknowledgements

Chapters in this Instant Insight are taken from the following sources:

Chapter 1 Microalgae: a unique source of poultry feed protein
Chapter taken from: Lei, X. G. (ed.), Seaweed and microalgae as alternative sources of protein, Burleigh Dodds Science Publishing, Cambridge, UK, 2021, (ISBN: 978 1 78676 620 5; www.bdspublishing.com)

Chapter 2 Use of protein from black solider flies in poultry feed
Chapter taken from: Casillas, A., Insects as alternative sources of protein for food and feed, Burleigh Dodds Science Publishing, Cambridge, UK, 2025, (ISBN: 978 1 80146 584 7; www.bdspublishing.com)

Chapter 3 Use of protein from yellow mealworms in poultry feed
Chapter taken from: Casillas, A., Insects as alternative sources of protein for food and feed, Burleigh Dodds Science Publishing, Cambridge, UK, 2025, (ISBN: 978 1 80146 584 7; www.bdspublishing.com)

Chapter 4 Alternative sources of protein for poultry nutrition: an overview
Chapter taken from: Applegate, T. J. (ed.), Advances in poultry nutrition, Burleigh Dodds Science Publishing, Cambridge, UK, 2024, (ISBN: 978 1 80146 731 5; www.bdspublishing.com)

Chapter 5 High protein corn fermentation products for poultry derived from corn ethanol production
Chapter taken from: Applegate, T. J. (ed.), Advances in poultry nutrition, Burleigh Dodds Science Publishing, Cambridge, UK, 2024, (ISBN: 978 1 80146 731 5; www.bdspublishing.com)

Chapter 1

Microalgae: a unique source of poultry feed protein

Sahil Kalia, Andrew D. Magnuson, Guanchen Liu and Xin Gen Lei, Cornell University, USA

1 Introduction

In an animal feed industry, corn and soybean meal (SBM) are the two most commonly used ingredients to meet energy and protein requirements of various food-producing species, in particular, simple-stomached animals, such as poultry and swine. In the United States, 60% of corn and 47% of soy are used to feed mainly chickens and pigs. Ultimately, this practice creates a direct competition between feed and food for these two staples (Olson, 2006). Consequently, the world may not afford to continue the feed use of corn and SBM at the current rate or scale without jeopardizing the global food security of humans with an increasing global population. Therefore, alternative feed ingredients, in particular, high protein feedstuff supplement, must be explored to sustain the current growth in animal production (Austic et al., 2013). One such alternative may be microalgae (Christaki et al., 2011). Microalgae have gained popularity recently as the third generation of feedstocks for biofuel production because many species of microalgae grow rapidly, synthesize high levels of lipids and sequester large amounts of CO_2 (Spolaore et al., 2006). Defatted algal biomass remained after the extraction of oils from the full-fatted microalgae is a rich source of proteins, amino acids and bioactive nutrients.

http://dx.doi.org/10.19103/AS.2021.0091.16

2 Nutrient composition of microalgae as alternative feed protein and bioactive nutrient sources

Both full-fatted and defatted microalgal biomass can be used as animal feed. Approximately, 30% of total microalgal biomass cultivated around the world is currently sold as animal feed (Vanthoor-Koopmans et al., 2013). Concentrations of proteins, vitamins and minerals found in microalgae biomass are usually nearly equivalent or higher than that found in most conventional feedstuffs such as corn, oil extracted dehulled SBM, wheat and fish meal (Lum et al., 2013). Indeed, the high protein contents and balanced amino acid profiles of various microalgae species are one of the main factors that favor their use in poultry diets (Kovac et al., 2013). Moreover, various types of defatted microalgae biomass generated after extraction of lipid during the biofuel production are manifested with higher protein concentrations than those of the full-fatted microalgae (Becker, 2007).

Table 1 lists concentrations of crude protein and essential amino acids in various sources of full- and de-fatted microalgae, along with comparisons with those in corn, SBM, wheat and fish meal. The crude protein concentrations in the listed microalgae species range from 14% in the full-fatted diatom *Staurosira* biomass to 76% in the blue-green *Spirulina* biomass, compared with 8.5%, 14%, 49% and 62% in corn, wheat, SBM and fish meal, respectively. The metabolizable energy in microalgae was reported as 1320 kcal/kg in the defatted *Staurosira* biomass, compared with 3560 kcal/kg, 3170 kcal/kg, 2425 kcal/kg and 2950 kcal/kg in corn, wheat, SBM and fish meal, respectively (Batal and Dale, 2010; Austic et al., 2013).

Concentrations of essential amino acids histidine (His), isoleucine (Ile), leucine (Leu), phenylalanine (Phy), threonine (Thr), tryptophan (Trp) and valine (Val) in various microalgae species range from 0.77%, 1.8%, 3.5%, 2.1%, 1.97%, 0.55% and 2.33% in the defatted *Nannochloropsis oceanica* to 2.0%, 3.8%, 8.8%, 5.0%, 4.8%, 3.4% and 5.5% in the green *Chlorella vulgaris*. In comparison, concentrations of these amino acids are 1.4%, 2.3%, 3.9%, 2.6%, 2.02%, 0.7% and 2.4% in SBM and 1.5%, 2.4%, 4.4%, 2.3%, 2.8%, 0.5% and 2.8% in fish meal, respectively. Concentrations of lysine (Lys) and methionine (Met), the two most limiting amino acids in corn-SBM-based diets for poultry, range from 2.4% and 0.67% in the defatted *N. oceanica* to 8.4% and 2.2% in *C. vulgaris*, compared with 3.0% and 0.76% in SBM and 4.7% and 1.7% in fish meal, respectively (Batal and Dale, 2010; NRC, 2012; Tao et al., 2018). Our laboratory has been evaluating nutrient compositions and feeding values of several full-fatted and defatted microalgae including *Staurosira* spp., *Desmodesmus* spp., *N. oceanica*, *Haematococcus pluvialis* and *Aurantiochytrium* biomass in the diets for broiler chickens and laying hens. Crude protein concentrations in those microalgae used in our laboratory range from 10% to 45%, which is 21-93%

of that in the SBM and 4.2-fold greater than that in the corn. Concentrations of ether extract (EE), acid detergent fiber, neutral detergent fiber and ash range from 14%, 0.70%, 14% and 45% in the defatted *Staurosira* biomass to 38%, 7.4%, 24% and 20% in the defatted *N. oceanica*. Comparatively, those concentrations are 1.0%, 10%, 16% and 6.0% in SBM and 1.9%, 4.2%, 9.7% and 1.2% in corn, respectively (NRC, 2012; Austic et al., 2013; Gatrell et al., 2015).

Table 2 shows fatty acid profiles of microalgae studied in our laboratory, in comparison with those of other oil ingredients (corn oil, soybean oil, fish oil, SBM). Concentrations of saturated fatty acids (SFA) were found to be highest in full-fatted *Staurosira* spp., intermediate in *N. oceanica* (CO 18 strain) and lowest in *Desmodesmus* spp. (CO 46 strain), respectively, compared with those (on basis of % total fatty acids) in corn oil, soybean oil, fish oil, and SBM. Concentrations of monounsaturated fatty acids (MUFA) were found to be highest in defatted *N. oceanica*, intermediate in *Aurantiochytrium* and lowest in defatted *Desmodesmus* spp. (CO46-LEA), respectively, compared those in fish oil, corn oil, soybean oil and SBM (Moir et al., 1995; Oladiji et al., 2009; Lee et al., 2015; Kim et al., 2016; Sun et al., 2020a,b). Concentrations of polyunsaturated fatty acid (PUFA) were found to be highest in defatted *Desmodesmus* spp., intermediate in defatted *N. oceanica* and lowest in full-fatted *Staurosira* spp., respectively, compared with those in SBM, corn oil, soybean oil and fish oil. Concentrations of total n-3 PUFA were found to be highest in *Aurantiochytrium*, intermediate in defatted *Desmodesmus* spp. (CO46-LEA) and lowest in full-fatted *Staurosira* spp., respectively, compared with those in fish oil, SBM, soybean oil and corn oil. Overall, n-6/n-3 PUFA ratios were much lower in various sources of the microalgal biomass than those in all feed oils except for fish oil (Moir et al., 1995; Oladiji et al., 2009; Lee et al., 2015; Kim et al., 2016).

Table 3 lists vitamin and mineral contents of various microalgae, in comparison with SBM and fish meal. Concentrations of riboflavin and niacin were found to be much higher in *Spirulina* spp. and higher in *Chlorella* spp. than in SBM. Concentrations of thiamine and α-tocopherol were much higher in *Spirulina* spp. and higher in *Chlorella* spp. than in SBM. Moreover, concentrations of folic acid were found to be highest in fish meal, intermediate in *Chlorella* and *Spirulina* spp. and lowest in SBM. Concentrations of calcium and phosphorus were found to be highest in fish meal, intermediate in *Staurosira* and lowest in *Desmodesmus*. Concentrations of sodium and potassium were found to be highest in *Staurosira*, intermediate in *Desmodesmus* and lowest in SBM and fish meal (Batal and Dale, 2010; Andrade et al., 2018). High concentrations of ash or sodium in the microalgal biomass may produce adverse effects on nutrient metabolism or animal health by causing feed refusal and excessive drinking water needs by birds.

Table 1 Protein and amino acid profile (%) of selected microalgae species and other ingredients used in poultry diets

Source	CP	Glu	Pro	Ile	Leu	Lys	Asp	Ala	Val	Arg	Gly	Phe	Thr	Ser	Tyr	His	Met	Trp	Cys	References
SBM	48.5	9.20	2.41	2.31	3.90	3.23	5.95	2.42	2.41	3.73	2.16	2.64	2.02	2.58	1.47	1.35	0.76	0.68	0.68	Sriperm et al., 2011
Corn	8.50	1.41	0.66	0.27	0.93	0.26	0.52	0.58	0.37	0.41	0.32	0.40	0.28	0.38	0.28	0.22	0.17	0.065	0.17	Sriperm et al., 2011
Wheat	13.5	–	–	0.69	1.0	0.40	–	–	0.69	0.60	–	0.78	0.35	–	–	0.17	0.25	0.18	0.30	Batal and Dale, 2010
Fish meal	62.0	–	–	2.40	4.40	4.70	–	–	2.80	3.65	–	2.28	2.75	–	–	1.52	1.70	0.50	0.50	Batal and Dale, 2010
Aurantiochytrium	10.3	1.04	0.28	0.35	0.59	0.47	1.12	0.54	0.47	0.42	0.41	0.36	0.42	0.42	0.27	0.20	2.28	0.13	0.12	Sun et al., 2020a
Chlorella pyrenoidosa	44.0	–	–	1.75	3.79	2.06	–	–	2.47	2.06	2.20	1.81	2.12	–	0.80	0.62	0.36	0.80	–	Combs, 1952
Chlorella vulgaris	58.0	11.6	4.8	3.8	8.8	8.4	9.0	7.9	5.5	6.4	5.8	5.0	4.8	4.1	3.4	2.0	2.2	2.1	1.4	Becker, 2007
Desmodesmus spp. defatted	31.2	2.93	2.73	1.10	2.29	1.61	2.69	2.27	1.59	1.45	1.72	1.34	1.26	1.10	1.01	0.50	0.48	0.43	0.33	Ekmay et al., 2014
Desmodesmus spp. (CO 46)	41.0	–	–	–	–	–	–	–	–	–	–	–	–	–	–	–	–	–	–	Sun et al., 2020b
Desmodesmus spp. (CO46-LEA)	39.0	3.44	3.84	1.51	2.97	2.20	3.08	2.94	2.00	1.84	2.06	1.83	1.44	1.12	1.27	0.65	0.65	0.24	0.55	Sun et al., 2020b
Haematococcus pluvialis defatted	28.5	–	2.20	0.13	2.70	2.49	–	4.63	1.31	1.87	–	2.06	–	–	2.03	0.56	0.78	0.21	0.07	Sun et al., 2018; Magnuson et al., 2018

Species																				Reference
H. pluvialis	26.8	–	1.59	1.16	3.22	2.59	–	4.89	1.67	0.87	–	2.45	–	–	1.27	0.42	0.87	0.17	0.07	Sun et al., 2018; Magnuson et al., 2018
Isochrysis spp.	27–45	4.60	2.40	1.80	3.90	2.50	4.20	3.20	2.40	2.50	2.60	3.80	2.40	2.20	3.80	0.90	1.40	0.60	1.40	Bandarra et al., 2003
Nannochloropsis oceanica	38.2	3.34	4.00	1.50	2.90	2.27	2.80	2.22	2.13	1.99	1.92	1.57	1.54	1.21	1.20	0.64	0.57	1.20	0.30	Gatrell et al., 2015
N. oceanica	45.1	4.13	2.38	1.83	3.51	2.37	3.53	–	2.33	2.23	2.36	2.05	1.97	1.63	1.52	0.77	0.67	0.55	0.42	Tao et al., 2018
N. oceanica (CO18 strain)	45.0	4.01	2.84	1.82	3.33	2.65	3.49	2.76	2.38	2.35	2.15	2.04	1.93	1.54	1.43	0.77	0.85	0.45	0.43	Unpublished report
Scenedesmus obliquus	55.0	10.7	3.6	5.6	7.3	5.6	8.4	9.0	6.0	7.1	7.1	4.8	5.1	3.8	3.2	2.1	1.5	0.3	0.6	Becker, 2007
Spirulina platensis	51.0–60.5	7.17	2.18	2.71	4.40	2.10	5.06	4.14	2.61	3.92	2.89	2.20	1.72	2.61	2.66	0.70	0.80	–	1.31	Tavernari et al., 2018
Spirulina spp.	76.0	6.12	1.72	2.55	3.88	2.10	4.58	3.21	2.94	–	2.32	–	1.97	–	–	–	1.07	–	0.42	Evans et al., 2015
Staurosira spp. defatted	19.1	1.81	0.65	0.78	1.33	0.83	1.88	1.09	0.98	0.93	0.96	0.86	0.88	0.76	0.57	0.30	0.33	1.18	0.32	Austic et al., 2013
Staurosira spp.	13.9	1.29	0.45	0.55	0.94	0.57	1.31	0.76	0.70	0.61	0.67	0.80	0.63	0.53	0.40	0.18	0.26	0.12	0.19	Ekmay et al., 2015

Ala, alanine; Arg, arginine; Asp, aspartic acid; CP, crude protein; Cys, cystine; Glu, glutamic acid; Gly, glycine; His, histidine; Ile, isoleucine; Leu, leucine; Lys, lysine; Met, methionine; Phe, phenylalanine; Pro, proline; SBM, soybean meal; Ser, serine; Thr, threonine; Trp, tryptophan; Tyr, tyrosine; Val, valine.

Table 2 Fatty acid composition of different microalgae species compared with conventional oils and soybean meal

Fatty acids	Staurosira spp.[1]	Defatted Desmodesmus spp.[1]	Defatted Nannochloropsis oceanica[1]	Desmodesmus spp. (CO 46 strain)[2]	Desmodesmus spp. (CO 46-LEA)[2]	N. oceanica (CO 18 strain)[3]	Aurantiochytrium[4]	Corn oil[5]	Fish oil[6]	Soybean oil[7]	Soybean meal[8,9,10]
						Content (% of total fatty acids)					
C14:0	8.31	1.21	7.44	0.59	0.48	4.77	2.19	0.1	9.70	0.10	0.14
C14:1	0.11	ND	0.12	ND	0.02	0.07	ND	ND	ND	ND	ND
C16:0	51.8	35.6	29.2	15.9	17.5	24.00	34.1	9.60	21.4	10.3	13.1
C16:1	35.7	1.59	25.2	3.08	2.66	21.0	ND	ND	15.2	5.20	0.13
C18:0	0.84	2.24	0.50	0.90	1.13	0.68	–	1.90	5.30	3.80	4.08
C18:1n9	0.71	19.3	12.1	9.03	8.19	5.79	–	29.0	15.9	22.8	16.2
C18:2n6	1.04	10.8	2.13	7.98	5.98	2.33	–	56.4	3.20	51.0	55.2
C18:3n3	ND	22.1	0.11	24.9	14.8	1.24	0.08	2.70	3.50	6.80	10.1
C18:3n6	0.24	1.64	0.42	1.38	1.22	0.28	–	ND	ND	ND	ND
C20:2n6	0.12	2.85	0.14	ND	ND	0.03	–	ND	ND	ND	ND
C20:4n6	0.41	ND	5.84	0.08	0.03	4.26	–	ND	ND	ND	ND
C20:5n3	0.49	0.41	16.4	0.47	ND	19.8	0.12	0.4	14.1	ND	ND
C22:5n3	ND	ND	ND	ND	ND	0.03	4.77	ND	1.80	ND	ND
C22:6n3	ND	ND	ND	0.10	0.05	0.09	23.01	ND	6.90	ND	ND
SFA	61.1	40.2	37.2	17.4	19.1	29.5	363	11.5	36.4	14.2	17.3
MUFA	36.6	21.9	37.7	12.1	10.9	26.8	–	29.0	31.0	28.0	16.3
PUFA	2.27	37.9	25.1	34.9	22.1	28.1	28.0	60.0	32.6	57.8	65.3

n-3	0.51	22.6	16.5	25.5	14.8	21.2	28.0	3.10	26.3	6.80	10.1
n-6	1.78	15.3	8.6	9.44	7.23	6.89	–	56.4	3.20	51.0	55.2
n-6/n-3	3.49	0.68	0.52	0.37	0.49	0.32	–	18.2	0.12	7.50	5.47
Ether extract (%)	30.1	1.10	1.50	–	–	–	–	6.0	9.20	20.0	1.51

MUFA, monounsaturated fatty acid; n-3, omega 3 fatty acid; n-6, omega 6 fatty acid; ND, not detected; PUFA, polyunsaturated fatty acid; SFA, saturated fatty acid. Source: [1]Kim et al., 2016; [2]Sun et al., 2020b; [3]Unpublished report; [4]Sun et al., 2020a; [5]Moir et al., 1995; [6]NRC, 2012; [7]Oladiji et al., 2009; [8]Lee et al., 2015; [9]Tang et al., 2012; [10]Batal and Dale, 2010.

Table 3 Vitamins and mineral concentrations of different microalgae species, soybean and fish meal

	Chlorella spp.[1]	Spirulina spp.[1]	Desmodesmus[2]	Staurosira[3]	Soybean meal[4]	Fish meal[4]
Thiamine (mg/kg)	17	24	–	–	1.7	0.20
Riboflavin (mg/kg)	43	37	–	–	2.6	4.8
Niacin (mg/kg)	238	128	–	–	20.9	55.0
Pantothenic acid (mg/kg)	11	0	–	–	13	8.8
Folic acid (µg/kg)	940	940	–	–	700	1000
Vitamin B12 (µg/kg)	1	0	–	–	0	150
α-Tocopherol (mg/kg)	15	50	–	–	3.3	5.7
Calcium (%)	–	–	0.33	2.8	0.31	4.8
Phosphorus (%)	–	–	0.65	0.76	0.72	3.0
Sodium (%)	–	–	3.2	3.9	0.04	0.68
Potassium (%)	–	–	0.89	1.7	2.1	0.96

Source: [1]Andrade et al., 2018; [2]Ekmay et al., 2014; [3]Austic et al., 2013; [4]Batal and Dale, 2010.

3 Effects and values of microalgae as a supplement in broiler diets

Table 4 summarizes the effects of supplemental microalgae on growth performance, health status and chicken meat quality in diets for broilers. Although we attempt to provide a comprehensive and representative summary, the list may not be exhaustive. In earlier studies, freshwater green microalgae *Chlorella* biomass was investigated as a potential source of high-quality protein and antioxidants for broiler chickens. In a 28-day experiment, Combs (1952) supplemented the broiler diet with *Chlorella pyrenoidosa* at 10% to replace SBM. The supplement improved body weight gain (BWG) and feed use efficiency. Grau and Klein (1957) demonstrated that the inclusion of protein-rich, sewage-grown *Chlorella* for 18 days up to 20% in broiler diets was well tolerated by the birds. Likewise, Lipstein and Hurwitz (1983) reported that the high crude protein content of *C. vulgaris* allowed the replacements of fish meal and SBM by 5% and 10% in the diet of broilers without any negative effects on their BWG and feed conversion ratio (FCR). In a 35-day experiment, Kang et al. (2013) recorded improved BWG and intestinal microbial population when broilers were fed with 1% *C. vulgaris*. Similar results were observed by An et al. (2016) that 0.5% of *Chlorella* supplementation for 35 days improved the BWG and FCR and elevated the level of serum immunoglobulin G (IgG) and immunoglobulin M (IgM). Further, Dlouha et al. (2008) observed an improved growth performance and higher selenium (Se) and glutathione peroxidase activity in the broiler breast meat after feeding the animals with Se-enriched *Chlorella* at 0.3 mg Se/kg of diet for 42 days.

After 41 days of feeding, Ross and Dominy (1990) found no difference in the growth performance of broilers supplemented with 1.5%, 3%, 6% and 12% of dehydrated green *Spirulina platensis*, compared with the controls. They concluded that *Spirulina* up to the level of 12% could be used to replace other protein sources. Evans et al. (2015) reported that replacement of SBM and corn with dried *Spirulina maxima* up to 16% in the broiler diets produced no negative effect on the performance. Feeding broilers 4% or 8% of dried *Spirulina* from 21 days to 36 days of age by Toyomizu et al. (2001) enhanced yellow pigmentation in the muscles, skin, fat and liver. A longer study of 12 weeks by Venkataraman et al. (1994) demonstrated no negative effects of dried *S. platensis* up to 14% in the diets on the growth performance, dressing percentage and histopathology of various organs. Moreover, another study conducted for 36 days by Shanmugapriya et al. (2015) showed an improved BWG, FCR and villus length in chickens fed a diet with 1% *S. platensis* biomass. Bonos et al. (2016) reported that dietary supplementation of *S. platensis* (5 g/kg or 10 g/kg) in broiler diets for 42 days enriched the meat with eicosapentaenoic acid (EPA) and docosahexaenoic acid (DHA), without negative impact on growth performance.

Table 4 Effects of dietary inclusions of different microalgae species on broiler performance, health and meat quality

Microalgae	Level in the diet	Main findings	References
Aurantiochytrium spp. Full-fatted	1.2, 2.4 and 4.9 g DHA/kg diet	Dose-dependent enrichments of DHA and decreased n-6/n-3 PUFA ratios and nonesterified fatty acid concentration in plasma, liver, muscle and adipose tissue. Upregulation of ACCα, SCD1, FASN, elongases 2 and 5, desaturases 1 and 2, ACS and CPT2 expression in adipose tissues	Tolba et al., 2019
Chlorella pyrenoidosa	10% as replacement of SBM	Increased growth performance and improved feed use efficiency	Combs, 1952
Chlorella spp.	10-30%	Level up to 20% improved growth rate and was well tolerated by broilers	Grau and Klein, 1957
Chlorella spp.	0.3 mg Se/kg of diet	Improved growth performance with increased glutathione peroxidase and decreased lipid peroxidation in breast meat	Dlouha et al., 2008
Chlorella vulgaris	5-10%	No adverse effects on the growth performance, FCR, carcass fat or liver size	Lipstein and Hurwitz, 1983
C. vulgaris	0.5%	No effects on BWG but increased the phygocytic activity of leucocytes	Kortbacek et al., 1994
C. vulgaris	1% as replacement of antibiotic growth promoters	Positive effects on BWG and IgG, IgM and Lactobacillus count in intestine	Kang et al., 2013
C. vulgaris	0.05%, 0.5% and 0.15%	Improved BWG, FCR and humoral immune response. No change in the meat quality, color, or pH	An et al., 2016
Desmodesmus spp. (DGM) defatted	15% as replacement of SBM and corn mixture	Improved FCR and altered the hepatic and muscle eIF4E, mTOR, S6 and pS6 protein levels	Ekmay et al., 2014
Haematococcus pluvalis	7, 36 and 179 mg astaxanthin/kg feed	No negative effects on BWG, feed intake, or FCR; increased astaxanthin and carotenoid concentrations in kidney, intestine and breast muscle.	Waldenstedt et al., 2003
H. pluvalis	10, 20, 40 and 80 mg/kg diet	Dose-dependent enrichments of astaxanthin, carotenoids, antioxidant capacity in plasma, liver, breast, thigh muscles; and improved meat quality	Sun et al., 2018
Nannochloropsis oceanica (DGA) defatted	2%, 4%, 8% and 16% as replacement of SBM and corn mixture	Enriched n-3 PUFA, EPA and DHA in plasma, liver, breast and thigh tissues. Decreased n-6/n-3 PUFA ratios and elevated the mRNA levels of fatty acid synthase and Δ-6 and Δ-9 desaturases. DGA inclusion at 2-8% had positive effects on BWG, FCR, relative organ weights and soluble inorganic phosphorus retention and excretion	Gatrell et al., 2015, 2017

Species	Inclusion level	Effects	Reference
N. oceanica defatted	10% as replacement of SBM and corn mixture	Increased EPA, DHA and antioxidant status in liver and muscles; decreased lipid peroxidation and cytochrome P450 (CYP2C23b, CYP2D6, CYP3A5 and CYP4V2) gene expression in liver and muscles	Tao et al., 2018
Schizochytrium JB5	0.1% and 0.2%	No adverse effects on growth performance or relative organ weights; improved meat quality with increased n-3 fatty acid and DHA content in breast meat	Yan and Kim, 2013
Schizochytrium limacinum	1% and 2% as replacement of SBM	Improvements in growth performance, carcass traits, serum composition, antioxidant capacity and EPA and DHA deposition in breast and thigh muscles	Long et al., 2018
Spirulina maxima	6%, 11%, 16% and 21%	Inclusion up to 16% had no adverse effects on broiler performance; improved digestibility of methionine	Evans et al., 2015
Spirulina platensis	1.5–12%	No adverse effects on BWG and FCR	Ross and Dominy, 1990
S. platensis	14–17% replacement of fish meal and groundnut cake	No negative effects on the BWG, meat quality, or histopathology of liver, heart and kidney	Venkataraman et al., 1994
S. platensis	0.5%, 1% and 1.5%	Improved BWG and FCR; increased villi height at inclusion level of 1%	Shanmugapriya et al., 2015
S. platensis	0.5% and 1%	No effects on growth performance or meat lipid peroxidation status; increased EPA and DHA content in thigh meat	Bonos et al., 2016
Spirulina spp.	4% and 8%	No negative effects on growth performance or relative weight of internal organs; increased yellow pigmentation in broiler flesh and liver	Toyomizu et al., 2001
Staurosira spp. (DFA) defatted	7.5% replacement of SBM or 7.5% and 10% replacements of SBM and corn mixture	Inclusion up to 7.5% for replacing a mixture of SBM and corn had no adverse effects on growth or plasma biochemistry; inclusion of 7.5% for replacing SBM improved performance in broilers with additions of essential amino acids (methionine, lysine, isoleucine, valine, tryptophan and threonine) to the diets	Austic et al., 2013

ACCα, acetyl-CoA carboxylase; ACS, acetyl-CoA synthetase; BWG, body weight gain; CPT2, carnitine palmitoyltransferase; DHA, docosahexaenoic acid; EPA, eicosapentaenoic acid; FASN, fatty acid synthase; FCR, feed conversion ratio; IgG, immunoglobulin G; IgM, immunoglobulin M; mRNA, messenger ribonucleic acid; mTOR, mammalian target of rapamycin; n-3, omega 3 fatty acid; n-6, omega 6 fatty acid; PUFA, polyunsaturated fatty acid; SBM, soybean meal; SCD1, stearoyl-CoA desaturase-1; Se, selenium.

In a 32-day experiment, Waldenstedt et al. (2003) evaluated the effects of supplemental astaxanthin-rich *Haematococcus pluvalis* microalgae meal (7 mg, 36 mg and 179 mg astaxanthin/kg feed) in broiler chickens infected with *Campylobacter jejuni*. They showed no difference in growth performance between different levels of supplementation, but tissue astaxanthin and carotenoid concentrations were significantly increased by higher levels of dietary microalgae. Yan and Kim (2013) reported increased DHA concentration and improved fatty acid profiles in broiler breast upon supplementation of 0.1–0.2% *Schizochytrium JB5*. Likewise, dietary inclusion of 1–2% of *Schizochytrium limacinum* for 42 days in broiler diets improved growth performance, decreased concentrations of serum cholesterol and low-density lipoprotein cholesterol (LDL-C) and increased concentrations of high-density lipoprotein cholesterol (HDL-C), antioxidants and fatty acids in breast and thigh muscle (Long et al., 2018).

Our laboratory has been testing the potential of full-fatted and/or defatted microalgal biomass of *Staurosira* spp., *Desmodesmus* spp., *N. oceanica* and *H. pluvialis* to replace conventional ingredients of corn and SBM (Table 4). Initially, Austic et al. (2013) reported that defatted diatom microalgae *Staurosira* biomass (DFA) could substitute for a mixture of 7.5% SBM and corn (1 part of SBM:3 parts of corn in a mixture) in a broiler diet. That level of replacement did not produce adverse effects on growth performance or plasma biomarkers. In contrast, the inclusion of DFA for replacing a mixture of 10% SBM and corn (1 part of SBM:4 parts of corn in a mixture) or 7.5% SBM impaired the growth performance of broilers. However, the lost performance associated with the 7.5% SBM replacement was partially restored by adding extra essential amino acids (methionine, lysine, isoleucine, valine, tryptophan and threonine) into the diet. A simple estimate is that the inclusion of 7.5% DFA in broiler diets to replace SBM could potentially spare over 2.4 million metric tons of soybean for human consumption annually (Austic et al., 2013).

Effects of feed ingredients on protein digestion and utilization in birds are often assessed by responses of plasma uric acid and amino acid concentrations (Milles and Featherston, 1974; Salter et al., 1974; Donsbough et al., 2010). The DFA supplementation in broiler diets decreased plasma uric acid concentrations, indicating an improved nitrogen utilization. In another study, Ekmay et al. (2014) evaluated the nutritional impacts of defatted green microalgae (*Desmodesmus* spp.) biomass (DGM) along with additions of protease and nonstarch polysaccharide degrading enzymes (NSPase) in diets for broilers for 42 days. Broilers fed with 15% DGM replacing the mixture of SBM and corn had improved growth performance and better gain/feed efficiency as compared with the control group. Intriguingly, supplementing NSPase and protease to the DGM-containing diets resulted in negative effects on growth performance. The DGM inclusion increased circulating plasma amino acid concentrations from

0.40 µmol/mL on day 21 to 2.20 µmol/mL on day 42 but did not alter plasma uric acid concentrations. The DGM inclusion further altered the hepatic and muscle levels of key proteins involved in the mammalian target of rapamycin (mTOR) pathway that regulates cell growth and nutrient metabolism.

Gatrell et al. (2015) investigated the effects of incorporating broiler diets with defatted green microalgae (*N. oceanica*) (DGA) for 42 days. Supplementing the diets for broilers with 0%, 2%, 4%, 8% and 16% DGA in replacing mixtures of corn and SBM resulted in linear increases in the concentrations of n-3 fatty acids, EPA and DHA, in the plasma, liver, breast and thigh. The 16% incorporation enriched the breast and thigh with EPA up to 0.54 and 0.88 mg/100 g tissue, DHA up to 0.79 and 1.04 mg/100 g tissue and n-3 PUFA up to 2.18 and 2.45 mg/100 g tissue, respectively. Broiler fed with 4% and 8% DGA displayed higher expression of Δ-6 and Δ-9 desaturase (genes involved in the biosynthesis of EPA and DHA) along with higher gene expression of fatty acid synthase than the control. The inclusions of DGA at 2%, 4% and 8% in the broiler diets did not create any adverse effect on BWG or FCR. In contrast, the highest level of inclusion (16%) led to adverse effects on the growth performance. The impairment was probably due to relative deficiency in the sulfur-containing amino acids methionine and cysteine, low digestibility of microalgal protein, or perhaps high nucleic acid content (ileal DNA retention was increased by DGA supplementation) (Gatrell et al., 2014; Gatrell et al., 2017). Meanwhile, a high sodium content of DGA also led to increased water intake of birds. However, the increasing DGA inclusions led to elevated soluble inorganic phosphorus retention and decreased inorganic phosphorus excretion. The decrease in the phosphorus excretion by the DGA inclusion may help minimize the environmental pollution of manure phosphorus excretion by broilers (Gatrell et al., 2017).

Tao et al. (2018) investigated the enrichments of n-3 PUFA in broilers by incorporating 10% defatted *N. oceanica* under various dietary corn oil, vitamin E and selenium concentrations. The incorporation of microalgae enriched the broiler breast and thigh with DHA up to 9.25 and 14.2 mg/100 g tissue, respectively. There was a significant downregulation in the hepatic expression of cytochrome P450 enzyme genes (CYP2C23b, CYP2D6, CYP3A5 and CYP4V2) involved in PUFA oxidation by the supplementation of defatted microalgae. In addition, supplementing 10% defatted *N. oceanica* in a corn-SBM basal diet increased the apparent retentions of dry matter and EE by 3.3% and 3.8%, respectively, compared with the controls (Sun et al., 2016). The supplementation decreased apparent ileal digestibilities of eight essential amino acids and six nonessential amino acids, ranging from 32% for isoleucine to 7% for glutamic acid (Sun et al., 2016).

Tolba et al. (2019) supplemented DHA-rich microalgae *Aurantiochytrium* to a corn-SBM basal diet for broilers for 42 days at 0, 1.2, 2.4 and 4.9 g DHA/ kg of diet, respectively, and found dose-dependent enrichments of DHA in the plasma, liver and muscle tissues. The DHA enrichments in the breast and thigh reached 82 and 96 mg/100 g tissue, respectively. As the United States Department of Agriculture (USDA) 2015-2020 Dietary Guidelines for Americans recommended daily intake of DHA is 250 mg of EPA + DHA (USDA, 2015; Cortinas et al., 2004), two servings (approximately 200 g) of the microalgae-supplementation derived chicken may meet a good portion of the suggested intake. Meanwhile, the microalgae incorporation upregulated the mRNA abundances of key genes involved in fatty acid de novo synthesis (ACCa, SCD1, FASN, elongase 2, elongase 5, desaturases 1 and desaturases 2) and oxidation (ACS and CPT2) in the adipose tissues but downregulated the same set of genes in the liver.

Sun et al. (2018) investigated the nutritional and metabolic effect of microalgal astaxanthin (AST) in diets for broilers under heat stress. Supplementation of AST from *H. pluvialis* at 80 mg/kg of diets enriched AST and carotenoids up to 2.16 mg/kg and 12.0 mg/kg, respectively, in the broiler breast muscle. The AST supplementation altered tissue redox status and intrinsic antioxidant enzyme activity and GSH concentrations. An increased meat color score and a decreased water holding capacity (WHC) were recorded in the breast and thighs. The AST supplementation decreased hepatic SFA, MUFA and PUFA concentrations, but did not affect the growth performance of chicks exposed to high room temperatures.

4 Effects and values of microalgae as a supplement in laying hen diets

Microalgal biomass incorporation into diets for laying hens is of particular interest due to the abundance of fat and the lipophilic nature of eggs for uptaking and depositing health-promoting compounds of microalgae. Table 5 outlines the effects of different microalgae supplementations in diets for laying hens on egg production and quality, along with related biochemical responses. Mariey et al. (2012) demonstrated that supplementing *S. platensis* at 0.10%, 0.15% and 0.20% for 24 weeks improved the egg production rate, egg mass, egg yolk color and egg yolk score and decreased yolk triglyceride and nonesterified free fatty acid concentrations. Selim et al. (2018) obtained similar results that *S. platensis* supplementation at 0.1%, 0.2% and 0.3% in hen diets from 38 weeks to 46 weeks of age enhanced egg production rate, egg mass, shell thickness and egg yolk score and decreased serum triglyceride concentrations and aspartate aminotransferase activities. Likewise, Zahroojian et al. (2013) reported significant increases in egg yolk color by supplementations of *S. platensis* at 1.5%, 2.0% and 2.5% of hen diets.

Walker et al. (2012) fed laying hens with AST-rich *H. pluvialis* at 0.49%, 1.47% and 2.94% in combination with tocomin for 8 weeks and noticed increased redness in color and concentration of tocopherol in the egg yolk. However, there was no change in egg production or egg yolk weight by the microalgae inclusion. The inclusions of *Chlorella* spp. at 30–120 g/kg of feed as a substitution of SBM in laying hen diets increased yellow pigmentation in the egg yolk whereas no changes occurred in egg production rate, egg weight, or shell quality (Lipstein et al., 1980). Grigorova (2005) reported that inclusions of 2% and 10% of dried biomass of freshwater *Chlorella* spp. in laying hen diets improved egg yolk pigmentation and egg production. Herber-McNeill and Elswyk (1998) observed increased redness and n-3 PUFA concentration in egg yolk by inclusions of *Schizochytrium* biomass at 2.4% and 4.8% in laying hen diets.

Inclusion of brown algae *Undaria pinnatifida* in laying hen diet at 0.5% enhanced egg production, organ weight and concentrations of blood albumin, triglycerides and total cholesterol (Choi et al., 2018). Carillo et al. (2012) fed birds with 2% of sardine oil in combination with 10% *Sargassum sinicola*, 10% *Enteromorpha* spp. or 10% *Macrocystis pyrifera* for 8 weeks. They found that supplementation of *S. sinicola* and *M. pyrifera* increased EPA whereas that of *Enteromorpha* spp. increased DHA concentrations in the egg yolk. Al-Harthi and El-Deek (2012) supplemented 3% and 6% of brown marine microalgae *Sargassum dentifebium* in hen diets from 23 to 42 weeks of age. The supplementations decreased cholesterol and triglyceride concentrations, but elevated high-density lipoproteins (HDL), carotenes and lutein plus zeaxanthin concentrations in egg yolk.

In our laboratory, Leng et al. (2014) supplemented 7.5% or 15% defatted *Staurosira* biomass (DFA) to replace SBM or a mixture of corn and SBM in laying hen diets for 8 weeks. Feeding DFA at 7.5% did not affect feed intake, egg weight, yolk weight, shell weight, shell thickness, egg specific gravity, egg-shell breaking strength or concentrations of yolk cholesterol and plasma uric acid. However, feeding DFA at 15% decreased feed intake, egg production and egg quality parameters as compared with the control. Inclusion of DFA increased redness and decreased lightness and yellowness of yolk color. This study revealed that 7.5% DFA could replace 7.5% SBM alone or a mixture of corn and SBM in diet of laying hens without adverse effects on health and egg quality. Ekmay et al. (2015) incorporated 25% defatted *Desmodesmus* and 11.7% full-fatted *Staurosira* in hen diets with or without added protease. Feeding hens with both microalgae for 8 weeks did not alter egg production, plasma concentrations of insulin, glutamine and uric acid, or plasma alkaline phosphatase activity, but reduced the plasma tartrate-resistant acid phosphatase activity (indicating status of bone health). The microalgae feeding also reduced plasma 3-methylhistidine concentration, improved the ileal amino acid digestibility, down-regulated duodenal *Lat1* and *Pept1* mRNA levels and up-regulated hepatic phospho-S6 ribosomal protein (PS6) production. This

Table 5 Effects of dietary inclusions of different microalgae on egg production and quality, health status and biochemical responses of laying hens

Microalgae	Level in the diet	Main findings	References
Chlorella spp.	3–12%	Egg yolks became a deep yellow color with no other effects.	Lipstein et al., 1980
Chlorella spp.	2.0–10%	Improved overall performance of layers; egg yolks became darker with algae inclusion.	Grigorova, 2005
Desmodesmus spp.	7.5–15%	No negative effects by 7.5% inclusion; reduced egg production and feed use efficiency by 15% inclusion; increased redness and decreased lightness and yellowness of yolk by all levels of inclusion.	Leng et al., 2014
Desmodesmus spp.	11.7%	Reduced plasma 3-methylhistidine concentration; improved ileal amino acid digestibility; decreased duodenal *Lat1* and *Pept1* mRNA levels; increased hepatic phospho-S6 ribosomal protein (PS6) levels.	Ekmay et al., 2015
Enteromorpha spp.	10%	Increased egg yolk content of n-3 fatty acid DHA.	Carillo et al., 2012
Haematococcus pluvialis	0.49–2.94%	Egg yolks becoming more orange-red in color with an increased tocopherol content.	Walker et al., 2012
H. pluvialis	0.2–0.8%	Microalgal astaxanthin improved the antioxidant capacity of eggs and the overall redox status of tissues.	Magnuson et al., 2018
Macrocystis pyrifera	10%	Increased total egg yolk n-3 fatty acid contents, particularly EPA.	Carillo et al., 2012
Nannochloropsis oceanica	0.1–1.0%	Enrichments of tissues and eggs with n-3 fatty acids EPA and DHA; darkening of egg yolk color.	Nitsan et al., 1999
N. oceanica	2.86–23%	Dose-dependent enrichments of EPA, DHA and n-3 fatty acids; altered mRNA expression of malic enzyme, fatty acid synthase, acetyl-CoA carboxylase, elongases 4 and 5, fatty acid desaturases 5, 6 and 9 in liver and muscles.	Manor et al., 2019
Sargassum dentifebium	3–6%	Elevation of plasma high-density lipoprotein; increased yolk palmitic acid, total carotene and lutein contents.	Al-Harthi and El-Deek, 2012
Sargassum sinicola	10%	Enriched egg yolks with EPA.	Carillo et al., 2012
Schizochytrium spp.	2.4–4.8%	Increased total egg yolk n-3 PUFA content; darken egg yolk color; no impact on taste by sensory panel.	Herber-McNeil and Elswyk, 1998

Species	Inclusion	Effect	Reference
Spirulina platensis	0.1–0.2%	Enhanced egg production, egg size and FCR.	Mariey et al., 2012
S. platensis	1.5–2.5%	Darken yolk color.	Zahroojian et al., 2013
S. platensis	0.1–0.3%	Improved egg production and egg quality; showing hepatoprotective activity.	Selim et al., 2018
Staurosira spp.	3.0–15%	Incremental enrichments of n-3 fatty acids in yolks; combined supplementations of flaxseed oil and microalgae elevated egg ALA content by 13- to 15-fold over the control (15.4 mg/egg). DHA content by twofold over the control (38 mg/egg) and EPA from undetectable to 15.4 mg/egg.	Kim et al., 2016
Undaria pinnatifida	0.5%	Increased liver and cecum weights; elevated blood concentrations of albumen, triglycerides and total cholesterol.	Choi et al., 2018

ALA, α-Linolenic acid; DHA, docosahexaenoic acid; EPA, eicosapentaenoic acid; FCR, feed conversion ratio; Lat 1, L-type amino-acid transporter; mRNA, messenger ribonucleic acid; n-3, omega 3 fatty acid; Pept 1, peptide transporter; PUFA, polyunsaturated fatty acid.

study revealed the feasibility of the inclusion of microalgae biomass at a high level as a source of protein in laying hens diet.

Kim et al. (2016) evaluated the potential of combining three different microalgae with flaxseed oil in enriching eggs with n-3 PUFA. The combined supplementations of 7.5% full-fatted *Staurosira*, 7.5% defatted *Desmodesmus*, 7.5 or 15% defatted *N. oceanica* and 3% flaxseed oil did not affect daily feed intake, plasma uric acid and inorganic phosphorus concentrations and alkaline phosphatase activity, egg production, egg weight, egg yolk weight or egg-shell thickness. The combined supplementations elevated egg yolk α-linolenic acid (ALA) content by 13- to 15-fold over the control (15.4 mg/egg), DHA content by twofold over the control (38 mg/egg) and EPA from undetectable to 15.4 mg/egg. Furthermore, the combined supplementations of flaxseed oil and the three microalgae enhanced egg yolk contents of PUFA by 23% and decreased egg yolk contents of SFA up to 29% and MUFA up to 18%, respectively. Subsequently, Magnuson et al. (2018) supplemented hen diet with AST-rich *H. pluvialis* at 10, 20, 40 and 80 mg of AST/kg of diet. The highest level of supplementation enriched AST and total carotenoids to 36.2 μg/g and 114 μg/g of egg yolk, respectively. The microalgal AST was absorbed and deposited into the tissues and eggs of layers, and consequently improved their redox status. The decreases in total malondialdehyde, an indicator of lipid peroxidation, and increases in oxygen radical absorbance capacity in both eggs and tissues of layers were AST dose-dependent. Total glutathione concentrations and activities of glutathione peroxidase and glutathione-S transferase in the liver of hens were down-regulated by the *H. pluvialis* supplementation, which might reflect a metabolic coordination between the intrinsic antioxidant defense and the enrichment of extrinsic phytochemical.

Manor et al. (2019) supplemented laying hen diets with defatted green microalgae (*N. oceanica*) (DGA) for 42 days at 0%, 2.86%, 5.75%, 11.5% and 23% to replace corn and SBM. They observed DGA dose-dependent enrichments of n-3 PUFA, EPA and DHA in the egg yolk and the liver, breast and thigh tissues. The 23% inclusion enriched the egg yolk and liver with EPA up to 0.62 mg/g and 0.05 mg/g, DHA up to 4.8 mg/g and 1.2 mg/g, n-3 PUFA up to 5.9 mg/g and 1.3 mg/g, respectively. Furthermore, the DGA inclusion altered the mRNA expression of malic enzyme, fatty acid synthase, acetyl-CoA carboxylase, elongases 4 and 5 and fatty acid desaturases 5, 6 and 9 in the liver and muscles of hens.

5 Potential of microalgae as a supplement in diets for other poultry species

A limited number of studies have tested the effects of microalgae inclusions in diets for other poultry species. Overall, the inclusions have produced positive effects on growth performance and product quality (Table 6). In a 35-day

Table 6 Effects of dietary inclusions of different microalgae in diets for poultry species other than broilers and layers

Microalgae	Level in the diet	Species	Main findings	References
Chlorella vulgaris	0%, 0.1%, 0.2% replacement of soybean meal	Pekin ducks	Increased final body weight, BWG, feed intake and yellowness, pH and water holding capacity of the duck meat.	Oh et al., 2015
Crypthecodinium cohnii	1.5% replacement of soybean meal	Muscovy ducks	Increased DHA concentrations in the breast.	Schiavone et al., 2007
Schizochytrium spp.	0.5%	Japanese quail	Increased DHA concentrations; decreased n-6/n-3 PUFA ratios and cholesterol concentrations in the egg yolk.	Gladkowski et al., 2014
Spirulina platensis	0.5%, 1.0%, 1.5%, 2.0%	Japanese quail	Increased yolk color and fertility of quail.	Ross and Dominy, 1990
S. platensis	0.1% or 0.2%	Japanese quail	Increased weight gain and improved FCR.	Yusuf et al., 2016
S. platensis	0%, 0.25%, 1%	Japanese quail	Increased BWG and feed intake; improved FCR and fertility; lowered serum cholesterol and free fatty acid concentrations.	Abouelezz, 2017
S. platensis	0%, 5%, 10% and 15%	Japanese quail	Increased yolk color, MUFA levels and total antioxidant capacity; decreased yolk lipid peroxidation.	Boiago et al., 2019

BWG, body weight gain; DHA, docosahexaenoic acid; FCR, feed conversion ratio; MUFA, monounsaturated fatty acid; n-3, omega 3 fatty acid; n-6, omega 6 fatty acid; PUFA, polyunsaturated fatty acid.

experiment, Ross and Dominy (1990) fed 1-week-old Japanese quail with 0%, 1.5%, 3.0%, 6.0% and 12.0% of *S. platensis*. The supplementations showed no effects on growth performance, egg production, or hatchability, but improved fertility and increased yolk color. In a 42-day experiment on Japanese quail, Boiago et al. (2019) obtained improvements in the egg yolk color, MUFA level and total antioxidant capacity by inclusions of *S. platensis* at 5%, 10% and 15% of diets. Similarly, Abouelezz (2017) observed improvements in BWG, FCR, fertility rate and reduced serum cholesterol concentrations by feeding Japanese quail with *S. platensis* at 0.25% and 1% in their diets.

Gladkowski et al. (2014) reported that supplementation of *Schizochytrium* (0.5%) in combination with flaxseed and *Saccharomyces cerevisiae* (0.03%) in diets for Japanese quail for 19 weeks enriched their eggs with n-3 PUFA, ALA and DHA up to 7.33%, 4.57% and 4.35% of total fatty acid, respectively. Oh et al. (2015) supplemented Pekin ducks with fermented *C. vulgaris* at 0%, 0.1% and 0.2% for 42 days and observed improved BWG, feed intake and meat quality parameters including meat color, shear force, pH and water holding capacity. Schiavone et al. (2007) enhanced DHA concentrations in breast meat of Muscovy ducks by a supplementation of *Crypthecodinium cohnii* at 5 g/kg of diet.

6 Conclusion and future trends

Past research has generated sufficient evidences to qualify selected types of microalgal biomass as viable and human health value-added alternative feed proteins and bioactive nutrient sources for poultry feeding. Future research is needed to improve nutrient digestibility and utilization of the biomass and the biomass-containing diets by additions of exogenous hydrolytic enzymes, limiting essential amino acids and deficient trace minerals in diets to complete and complement the nutritional values of microalgae. Minimizing sodium concentrations in defatted marine microalgae will prevent resultant side effects of high salinity on feed and water intakes by birds. Special native or engineered microalgae strains should be developed to make the biomass a dual source of high-quality protein and bioactive nutrients including n-3 PUFA to various types of poultry. The threshold of microalgae supplementation in various poultry diets requires systematic evaluation. It remains unclear to us if the above described adverse effects of high microalgae inclusions such as the 16% DGA in broiler diets were attributed to the high salt content of seawater grown algae, high nucleic acid content in the unicellular diatom, or high resistance of microalgal cell wall to digestion. Identifying toxic and harmful microalgae strains and excluding their usage in poultry and other animal feeds should not be neglected.

The United States broiler and egg industries consume approximately 22 million tons of SBM annually (United Soybean Board, 2012). An estimated replacement of 5-20% of corn and SBM by microalgae in poultry diets would spare 575-2300 thousand tons of corn and SBM annually for human consumption (Bruinsma, 2003). Climate change, energy sustainability and food security are among the most important global challenges during the twenty-first century (Greene et al., 2016). The abundances of proteins, essential amino acids, n-3 PUFA and other essential micronutrients in microalgae will enable the development of not only alternative feed protein but also massive algal food industry toward commercialization of healthy and functional foods (Koyande et al., 2019). A dual production of algal food and fuel products at large scales would help the world to cope with the global challenges of food, energy, and water demands and climate changes (Efroymson et al., 2016).

7 Where to look for further information

Department of Animal Science, Cornell University (https://cals.cornell.edu/animal-science/research). The Lei Lab has been performing basic and applied research with a large group of world leading scientists to develop microalgae-based alternative feed and food protein and bioactive supplements to improve animal production, human health, and environmental sustainability to mitigate climate change.

- Marine AlGae Industrialization Consortium (MAGIC) (https://www.algae consortium.com/magic/).
- Department of Poultry Science at University of Arkansas (https://poultry -science.uark.edu/).
- United State Department of Agriculture (https://www.usda.gov/).

8 References

Abouelezz, F. M. K. (2017). Evaluation of *Spirulina* algae (*Spirulina platensis*) as a feed supplement for Japanese quail: nutritional effects on growth performance, egg production, egg quality, blood metabolites, sperm-egg penetration and fertility, *Poult. Sci. J.* 37(3): 707-719.

Al-Harthi, M. A. and El-Deek, A. A. (2012). Effect of different dietary concentrations of brown Marine algae (*Sargassum dentifebium*) prepared by different methods on plasma and yolk lipid profiles, yolk total carotene and lutein plus zeaxanthin of laying hens, *Ital. J. Anim. Sci.* 11(4). doi: 10.4081/ijas.2012.e64.

An, B.-K., Kim, K.-E., Jeon, J. Y. and Lee, K. W. (2016). Effect of dried *Chlorella vulgaris* and *Chlorella* growth factor on growth performance, meat qualities and humoral immune responses in broiler chickens, *Springerplus* 5(1): 718.

Andrade, L. M., Andrade, C. J., Dias, M., Nascimento, C. A. and Mendes, M. A. (2018). *Chlorella* and *Spirulina* microalgae as sources of functional foods, nutraceuticals, and food supplements; an overview, *MOJ Food Process. Technol.* 6(1): 5-58.

Austic, R. E., Mustafa, A., Jung, B., Gatrell, S. and Lei, X. G. (2013). Potential and limitation of a new defatted diatom microalgal biomass in replacing soybean meal and corn in diets for broiler chickens, *J. Agric. Food Chem.* 61(30): 7341-7348.

Bandarra, N. M., Pereira, P. A., Batista, I. and Vilela, M. H. (2003). Fatty acids, sterols and α-tocopherol in *Isochrysis galbana*, *J. Food Lipids* 10(1): 25–34.

Batal, A. and Dale, N. (2010). Ingredient analysis table: 2011 addition, *Feedstuffs* 2010: 16–17.

Becker, E. W. (2007). Micro-algae as a source of protein, *Biotechnol. Adv.* 25(2): 207–210.

Boiago, M. M., Dilkin, J. D., Kolm, M. A., Barreta, M., Souza, C. F., Baldissera, M. D., Dos Santos, I. D., Wagner, R., Tavernari, F. C., da Silva, M. L. B., Zampar, A., Stivanin, T. E. and Da Silva, A. S. (2019). *Spirulina platensis* in Japanese quail feeding alters fatty acid profiles and improves egg quality: benefits to consumers, *J. Food Biochem.* 43(7): e12860 doi: 10.1111/jfbc.12860.

Bonos, E., Kasapidou, E., Kargopoulos, A., Karampampas, A., Christaki, E., Florou-Paneri, P. and Nikolakakis, I. (2016). *Spirulina* as a functional ingredient in broiler chicken diets, *S. Afr. J. Anim. Sci.* 46(1): 94–102.

Bruinsma, J. (2003). *World Agriculture: Towards 2015/2030; an FAO Perspective*, Earthscan Publications, London, UK.

Carillo, S., Rios, V. H., Calvo, C., Carranco, M. E., Casas, M. and Pérez-Gil, F. (2012). n-3 fatty acid content in eggs laid by hens fed with marine algae and sardine oil and stored at different times and temperatures, *J. Appl. Phycol.* 24(3): 593–599.

Choi, Y., Lee, E. C., Na, Y. and Lee, S. R. (2018). Effects of dietary supplementation with fermented and non-fermented brown algae by-products on laying performance, egg quality, and blood profile in laying hens, *Asian-Australas. J. Anim. Sci.* 31(10): 1654–1659.

Christaki, E., Florou-Paneri, P. and Bonos, E. (2011). Microalgae: a novel ingredient in nutrition, *Int. J. Food Sci. Nutr.* 62(8): 794–799.

Combs, G. F. (1952). Algae (*Chlorella*) as a source of nutrients for the chick, *Science* 116(3017): 453–454.

Cortinas, L., Villaverde, C., Galobart, J., Baucells, M. D., Codony, R. and Barroeta, A. C. (2004). Fatty acid content in chicken thigh and breast as affected by dietary polyunsaturation level, *Poult. Sci.* 83(7): 1155–1164.

Dlouha, G., Sevcikova, S., Dokoupilova, A., Zita, L., Heindl, J. and Skřivan, M. (2008). Effect of dietary selenium sources on growth performance, breast muscle selenium, glutathione peroxidase activity and oxidative stability in broilers, *Czech J. Anim. Sci.* 53(6): 265–269.

Donsbough, A. L., Powell, S., Waguespac, A., Bidner, T. D. and Southern, L. L. (2010). Uric acid, urea, and ammonia concentrations in serum and uric acid concentration in excreta as indicators of amino acid utilization in diets for broilers, *Poult. Sci.* 89(2): 287–294.

Efroymson, R. A., Dale, V. H. and Langholtz, M. H. (2016). Socioeconomic indicators for sustainable design and commercial development of algal biofuel systems, *Glob. Change. Biol. Bioenergy* 9(6): 1005–1023.

Ekmay, R., Gatrell, S., Lum, K., Kim, J. and Lei, X. G. (2014). Nutritional and metabolic impacts of a defatted green marine microalgal (*Desmodesmus* sp.) biomass in diets for weanling pigs and broiler chickens, *J. Agric. Food Chem.* 62(40): 9783–9791.

Ekmay, R. D., Chou, K., Magnuson, A. and Lei, X. G. (2015). Continual feeding of two types of microalgal biomass affected protein digestion and metabolism in laying hens, *J. Anim. Sci.* 93(1): 287–297.

Evans, A. M., Smith, D. L. and Moritz, J. S. (2015). Effects of algae incorporation into broiler starter diet formulations on nutrient digestibility and 3 to 21 d bird performance, *J. Appl. Poult. Res.* 24(2): 206–214.

Gatrell, S., Lum, K., Kim, J. and Lei, X. G. (2014). Nonruminant nutrition symposium: potential of defatted microalgae from the biofuel industry as an ingredient to replace corn and soybean meal in swine and poultry diets, *J. Anim. Sci.* 92(4): 1306–1314.

Gatrell, S. K., Derksen, T. J., O'Neil, E. V. and Lei, X. G. (2017). A new type of defatted green microalgae exerts dose-dependent nutritional, metabolic, and environmental impacts in broiler chicks, *J. Appl. Poult. Res.* 26(3): 358–366.

Gatrell, S. K., Kim, J., Derksen, T. J., O'Neil, E. V. and Lei, X. G. (2015). Creating omega-3 fatty-acid-enriched chicken using defatted green microalgal biomass, *J. Agric. Food Chem.* 63(42): 9315–9322.

Gładkowski, W., Kiełbowicz, G., Chojnacka, A., Bobak, Ł., Spychaj, R., Dobrzański, Z., Trziszka, T. and Wawrzeńczyk, C. (2014). The effect of feed supplementation with dietary sources of n-3 polyunsaturated fatty acids, flaxseed and algae *Schizochytrium* sp., on their incorporation into lipid fractions of Japanese quail eggs, *Int. J. Food Sci. Technol.* 49(8): 1876–1885.

Grau, C. R. and Klein, N. W. (1957). Sewage-grown algae as a feedstuff for chicks, *Poult. Sci.* 36(5): 1046–1051.

Greene, C. H., Huntley, M. E., Archibald, I., Gerber, L. N., Sills, D. L., Granados, J., Tester, J., Beal, C., Walsh, M., Bidigare, R., Brown, S., Cochlan, W., Johnson, Z., Lei, X. G., Machesky, S., Redalje, D., Richardon, R., Kiron, V. and Corless, V. (2016). Marine microalgae: climate, energy, and food security from the sea, *Oceanography* 29(4): 10–15.

Grigorova, S. (2005). Dry biomass of freshwater algae of chlorella genus in the combined forages for laying hens, *J. Cent. Eur. Agric.* 6(4): 625–630.

Herber-McNeill, S. M. and Elswyk, M. E. Van (1998). Dietary marine algae maintain egg consumer acceptability while enhancing yolk color, *Poult. Sci.* 77(3): 493–496.

Kang, H. K., Salim, H. M., Akter, N., Kim, D. W., Kim, J. H., Bang, H. T., Kim, M. J., Na, J. C., Hwangbo, J., Choi, H. C. and Suh, O. S. (2013). Effect of various forms of dietary *Chlorella* supplementation on growth performance, immune characteristics, and intestinal microflora population of broiler chickens, *J. Appl. Poult. Res.* 22(1): 100–108.

Kim, J., Magnuson, A., Tao, L., Barcus, M. and Lei, X. G. (2016). Potential of combining flaxseed oil and microalgal biomass in producing eggs-enriched with n-3 fatty acids for meeting human needs, *Algal Res.* 17: 31–37.

Kotrbacek, V., Halouzka, R., Jurajda, V., Knotkova, Z. and Filka, J. (1994). Increased immune response in broilers after administration of natural food supplements, *Vet. Med. (Praha)* 39(6): 321–328.

Kovac, D. J., Simeunovic, J. B., Babic, O. B., Misan, A. C. and Milovanovic, I. L. (2013). Algae in food and feed, *Food Feed Res.* 40: 21–31.

Koyande, A. K., Chew, K. W., Rambabu, K., Tao, Y., Chu, D. T. and Show, P. (2019). Microalgae: a potential alternative to health supplementation for humans, *Food Sci. Hum. Well* 8(1): 16–24.

Lee, J. W., Kil, D. Y., Keever, B. D., Killefer, J., McKeith, F. K., Sulabo, R. C. and Stein, H. H. (2015). Carcass fat quality of pigs is not improved by adding corn germ, beef tallow, palm kernel oil, or glycerol to finishing diets containing distillers dried grains with solubles, *J. Anim. Sci.* 91(5): 2426–2437.

Leng, X. J., Hsu, K. N., Austic, R. E. and Lei, X. G. (2014). Effect of dietary defatted diatom biomass on egg production and quality of laying hens, *J. Anim. Sci. Biotechnol.* 5(1): 3. doi: 10.1186/2049-1891-5-3:Artn310.1186/2049-

Lipstein, B. and Hurwitz, S. (1983). The nutritional value of sewage-grown samples of *Chlorella* and *Micractinium* in broiler diets, *Poult. Sci.* 62(7): 1254–1260.

Lipstein, B., Hurwitz, S. and Bornstein, S. (1980). The nutritional value of algae for poultry. Dried *Chlorella* in layer diets, *Br. Poult. Sci.* 21(1): 23–27.

Long, S. F., Kang, S., Wang, Q. Q., Xu, Y. T., Pan, L., Hu, J. X., Li, M. and Piao, X. S. (2018). Dietary supplementation with DHA-rich microalgae improves performance, serum composition, carcass trait, antioxidant status, and fatty acid profile of broilers, *Poult. Sci.* 97(6): 1881–1890.

Lum, K. K., Kim, K. and Lei, X. G. (2013). Dual potential of microalgae as a sustainable feedstock and animal feed, *J. Anim. Sci. Technol.* 4: 1–7.

Magnuson, A. D., Sun, T., Yin, R., Liu, G., Tolba, S., Shinde, S. and Lei, X. G. (2018). Supplemental microalgal astaxanthin produced coordinated changes in intrinsic antioxidant systems of layer hens exposed to heat stress, *Algal Res.* 33: 84–90.

Manor, M. L., Derksen, T. J., Magnuson, A. D., Raza, F. and Lei, X. G. (2019). Inclusion of dietary defatted microalgae dose-dependently enriches omega-3 fatty acids in egg yolk and tissues of laying hens, *J. Nutr.* 149(6): 942–950.

Mariey, Y. A., Samak, H. R. and Ibrahem, M. A. (2012). Effect of using *Spirulina platensis* algae as a feed additive for poultry diets. 1. Productive and reproductive performances of local laying hens. *Egypt, Poult. Sci.* 32: 201–215.

Milles, R. D. and Featherston, W. R. (1974). Uric acid excretion as an indicator of the amino acid requirement of chicks, *Proc. Soc. Exp. Biol. Med.* 145(2): 686–689.

Moir, A. M. B., Park, B. S. and Zammit, V. A. (1995). Quantification in vivo of the effects of different types of dietary fat on theloci of control involved in hepatic triacylglycerol secretion, *Biochem. J.* 308(2): 537–542.

National Research Council (2012). *Nutrient Requirements of Swine* (11th rev. edn.), The National Academies Press, Washington, DC.

Nitsan, Z., Mokady, S. and Sukenik, A. (1999). Enrichment of poultry products with omega-3 512 fatty acids by dietary supplementation with the alga *Nannochloropsis* and mantur oil, *J. Agric. Food Chem.* 47(12): 5127–5132.

Oh, S. T., Zheng, L., Kwon, H. J., Choo, Y. K., Lee, K. W., Kang, C. W. and An, B. K. (2015). Effects of dietary fermented chlorella vulgaris (cbt(®)) on growth performance, relative organ weights, cecal microflora, tibia bone characteristics, and meat qualities in Pekin ducks, *Asian-Australas. J. Anim. Sci.* 28(1): 95–101.

Oladiji, A. T., Shoremekun, K. L. and Yakubu, M. T. (2009). Physicochemical properties of the oil from the fruit of *Blighia sapida* and toxicological evaluation of the oil-based diet in Wistar rats, *J. Med. Food* 12(5): 1127–1135.

Olson, R. D. (2006). *Below Cost Feed Crops: An Indirect Subsidy for Industrial Animal Factories*, Institute for Agriculture and Trade Policy, Minneapolis, MN.

Ross, E. and Dominy, W. (1990). The nutritional value of dehydrated, blue-green algae (*Spirulina plantensis*) for poultry, *Poult. Sci.* 69(5): 794–800.

Salter, D. N., Coates, M. E. and Hewitt, D. (1974). The utilization of protein and excretion of uric acid in germ-free and conventional chicks, *Br. J. Nutr.* 31(3): 307–318.

Schiavone, A., Chiarini, R., Marzoni, M., Castillo, A., Tassone, S. and Romboli, I. (2007). Breast meat traits of Muscovy ducks fed on a microalga (Crypthecodinium cohnii) meal supplemented diet, *Br. Poult. Sci.* 48(5): 573–579.

Selim, S., Hussein, E. and Abou-Elkhair, R. (2018). Effect of *Spirulina platensis* as a feed additive on laying performance, egg quality and hepatoprotective activity of laying hens, *Eur. Poult. Sci.* 82: 14–24.

Shanmugapriya, B., Babu, S. S., Hariharan, T., Sivaneswaran, S. and Anusha, M. B. (2015). Dietary administration of *Spirulina platensis* as probiotics on growth performance and histopathology in broiler chicks, *Int. J. Recent Sci. Res.* 6: 2650-2653.

Spolaore, P., Cassana, C. J., Duran, E. and Isamberta, A. (2006). Commercial applications of microalgae, *J. Biosci. Bioeng.* 101(2): 87-96.

Sriperm, N., Pesti, G. M. and Tillman, P. B. (2011). Evaluation of the fixed nitrogen-to-protein (N:P) conversion factor (6.25) versus ingredient specific N:P conversion factors in feedstuffs, *J. Sci. Food Agric.* 91(7): 1182-1186.

Sun, T., Magnuson, A. and Lei, X. G. (2016). New perspective of nutrient digestibility and retention in diets containing defatted microalgae, The 78th Cornell Nutrition Conference, East Syracuse, New York.

Sun, T., Tolba, S., Magnuson, A., Liu, G. and Lei, X. G. (2020a). Supplemental dietary DHA-rich microalgae affected growth performance, health status and meat quality of broiler chicks, Science Association Annual, Poultry Meeting, Virtual.

Sun, T., Wyman, B., Liu, G. and Lei, X. G. (2020b). Dietary supplemental full-fatted and defatted microalgae *Desmodesmus* sp. exerted similar impacts on growth performance and gut health of broiler chicks, Science Association Annual, Poultry Meeting, Virtual.

Sun, T., Yin, R., Magnuson, A. D., Tolba, S. A., Liu, G. and Lei, X. G. (2018). Dose-dependent enrichments and improved redox status in tissues of broiler chicks under heat stress by dietary supplemental microalgal astaxanthin, *J. Agric. Food Chem.* 66(22): 5521-5530.

Tang, J. W., Sun, H., Yao, X. H., Wu, Y. F., Wang, X. and Feng, J. (2012). Effects of replacement of soybean meal by fermented cottonseed meal on growth performance, serum biochemical parameters and immune function of yellow-feathered broilers, *Asian-Australas J. Anim. Sci.* 25(3): 393-400.

Tao, L., Sun, T., Magnuson, A. D., Qamar, T. R. and Lei, X. G. (2018). Defatted microalgae-mediated enrichment of n–3 polyunsaturated fatty acids in chicken muscle is not affected by dietary selenium, vitamin E, or corn oil, *J. Nutr.* 148(10): 1547–1555.

Tavernari, F. C., Roza, L. F., Surek, D., Sordi, C., Silva, M. L. B. D., Albino, L. F. T., Migliorini, M. J., Paiano, D. and Boiago, M. M. (2018). Apparent metabolisable energy and amino acid digestibility of microalgae *Spirulina platensis* as an ingredient in broiler chicken diets, *Br. Poult. Sci.* 59(5): 562-567.

Tolba, S. A., Sun, T., Magnuson, A. D., Liu, G. C., Abdel-Razik, W. M., El-Gamal, M. F. and Lei, X. G. (2019). Supplemental docosahexaenoic-acid-enriched microalgae affected fatty acid and metabolic profiles and related gene expression in several tissues of broiler chicks, *J. Agric. Food Chem.* 67(23): 6497-6507.

Toyomizu, M., Sato, K., Taroda, H., Kato, T. and Akiba, Y. (2001). Effects of dietary *Spirulina* on meat colour in muscle of broiler chickens, *Br. Poult. Sci.* 42(2): 197-202.

United Soybean Board (2012). *Animal Agriculture Economic Analysis: National, 2001-2011. A Report for United Soybean Board*, Agralytica, Inc., Alexandria, VA.

US Department of Agriculture (2015-2020). Dietary guidelines for Americans. https://www.fns.usda.gov/cnpp/dietary-guidelines-americans, accessed November 08, 2020.

Vanthoor-Koopmans, M., Wijffels, R. H., Barbosa, M. J. and Eppink, M. H. (2013). Biorefinery of microalgae for food and fuel, *Bioresour. Technol.* 135: 142-149.

Venkataraman, L. V., Somasekaran, T. and Becker, E. W. (1994). Replacement value of blue-green alga (*Spirulina platensis*) for fishmeal and a vitamin-mineral premix for broiler chicks, *Br. Poult. Sci.* 35(3): 373-381.

Waldenstedt, L., Inborr, J., Hansson, I. and Elwinger, K. (2003). Effects of astaxanthin-rich algal meal (*Haematococcus pluvalis*) on growth performance, caecal campylobacter and clostridial counts and tissue astaxanthin concentration of broiler chickens, *Anim. Feed Sci. Technol.* 108(1-4): 119-132.

Walker, L. A., Wang, T., Xin, H. and Dodle, D. (2012). Supplementation of laying-hen feed with palm tocos and algae astaxanthin for egg yolk nutrient enrichment, *J. Agric. Food Chem.* 60(8): 1989-1999.

Yan, L. and Kim, I. H. (2013). Effects of dietary ω-3 fatty acid-enriched microalgae supplementation on growth performance, blood profiles, meat quality, and fatty acid composition of meat in broilers, *J. Appl. Anim. Res.* 4: 392-397.

Yusuf, M. S., Hassan, M. A., Abdel-Daim, M. M., Nabtiti, A. S. E., Ahmed, A. M., Moawed, S. A., El-Sayed, A. K. and Cui, H. (2016). Value added by Spirulina platensis in two different diets on growth performance, gut microbiota, and meat quality of Japanese quails, *Vet. World* 9(11): 1287-1293.

Zahroojian, N., Moravej, H. and Shivazad, M. (2013). Effects of dietary marine algae (*Spirulina platensis*) on egg quality and production performance of laying hens, *J. Agr. Sci. Technol.* 15: 1353-1136.

Chapter 2

Black soldier fly meal: an alternative protein source for pigs

S. Struthers, and J. G. M. Houdijk, Scotland's Rural College (SRUC), UK; and H. N. Hall, Anpario plc, UK

1 Introduction

The rise in the global human population, in conjunction with higher standards of living, has increased the demand for animal protein for human consumption. The overall demand for protein annually is estimated to be 202 million tonnes but is expected to reach 435 million tonnes by 2050 (Boland et al., 2013; Henchion et al., 2017). It is predicted that a large proportion of this demand will be for pork and poultry products (Kim et al., 2019), given that these are the more efficient land-based, farmed animals for converting feed protein into animal protein, but this would result in increased pressure on sustainable protein sourcing of their nutrition.

Pork is currently the second most consumed protein (globally), behind poultry. It is projected to grow by 11% to 129 million tonnes by 2032 (OECD/ FAO, 2023), and as a versatile and lean meat, pork can be highly efficient to produce. Currently, most of the protein requirement in pig diets is fulfilled by plant sources, predominantly soya bean meal (SBM) (Boland et al., 2013; Kim et al., 2019; Alagappan et al., 2022). However, using soya products in

http://dx.doi.org/10.19103/AS.2024.0139.05

livestock feed is largely considered unsustainable due to increasing demand for SBM without an increase in supply, which drives prices upwards, but also the significant climate impact of land use change especially for meeting this additional SBM demand, as well as the carbon footprint of the associated transport distances (Ffoulkes et al., 2021; Gupta et al., 2021). Fish meal is no longer routinely included in pig diets due to unfavourable raw material costs, as the majority of FM produced is used in aquaculture, but it is known to be a valuable and well balanced protein for young pigs especially in creep and post-weaning diets. In summary, the increased protein requirements for human consumption drives a need for alternative sustainable, protein-rich, raw materials to feed livestock.

Insects, especially as it can be argued that using pigs as bio-converters to produce high-value protein products from nutritional resources that are not suitable for direct consumption by humans, will assist in the overall sustainability of animal protein production. Since humans can readily utilise soya as a source of protein, following appropriate food preparation, more protein sources are required which do not compete with human edible sources. These alternative proteins should be suitable for feeding, cost effective and have improved sustainability credentials compared to traditional sources such as SBM.

Black soldier fly (BSF; *Hermetia illucens*) is a promising alternative protein for livestock diets. BSF can be reared on a variety of co-product streams from the worldwide food and drink industries, as well as from household and retail food surplus. This allows for bioconversion of a wide array of organic substrates, converting this material from a relatively low protein value, in terms of variable quality, into a source of much less variable, high-quality nutrients for human consumption (Spranghers et al., 2017). BSF reared on organic substrates is rich in crude protein and fats and has a nutritional profile comparable to that of SBM and FM in terms of amino acid profile. Therefore, it offers excellent soya-replacement potential (Spranghers et al., 2017; Heuel et al., 2022).

2 Nutritional composition of black soldier fly

The nutritional composition and dietary value of BSF have been studied and reviewed in depth in several papers (Barragan-Fonseca et al., 2017; Spranghers et al., 2017; Veldkamp and Vernooij, 2021; Hawkey and Hall, 2023). However, papers reviewing the commercial replacement of traditional protein sources in pig diets in a commercial setting are still relatively lacking. As protein makes up the largest component of BSF on a dry matter basis, followed by fat (Barragan-Fonseca et al., 2017), it would also require post-harvest processing to best fit the commercial diet of pigs and more closely resemble current protein sources; this would then enable the most straightforward nutritional comparison and exchange of the tradition protein source for BSF in the diet. The average crude protein and fat content in BSF are 40% and 26% on a dry matter basis, respectively; however, these values can vary based on the feed substrate and insect developmental stage at the time of harvest (Barragan-Fonseca et al., 2017; Hawkey and Hall, 2023).

BSF's amino acid profile is comparable to that of SBM and fishmeal (FM) (Hawkey and Hall, 2023). The amino acid content of BSF also does not vary much between studies; however, specific amino acid levels can differ due to feed substrate (Barragan-Fonseca et al., 2017; Spranghers et al., 2017). BSF protein contains 10 essential amino acids and has higher levels of histidine, methionine, and tryptophan when compared to SBM (Hawkey and Hall, 2023).

Saturated fatty acids comprise most of the total fat content in BSF (Barragan-Fonseca et al., 2017), with the major fatty acids being palmitic, oleic, and linoleic acids. BSF also has a high level of lauric acid, a medium-chain saturated fatty acid, which has been reported to have anti-microbial and anti-inflammatory properties (Spranghers et al., 2017).

The mineral content of BSF is generally higher than other insect species used as feed (Finke, 2013). BSF larvae, in particular, have high calcium content due to their mineralised exoskeleton (Finke, 2013); however, some of this calcium is bound to chitin, the structural carbohydrate found in insects and shellfish, making its bioavailability to the pig limited as, like most animals, it does not produce chitinases endogenously (Rathore and Gupta, 2015).

Chitin is a long-chain polysaccharide found within the exoskeleton of insects and shellfish, including prawns and lobsters (Komi et al., 2018). Chitin is resistant to degradation by the animal in the digestive tract; however, it has been shown to promote hindgut fermentation of short-chain fatty acids and have immunomodulatory effects (Komi et al., 2018; Yu et al., 2020b). This might suggest that chitin can exert prebiotic properties. Chitin has been shown to have value once extracted from BSF for use in industrial processes (bioplastics) and animal diets but at lower levels. When used to modulate gut health, it has the potential to bind endotoxins and mycotoxins.

When estimating the crude protein content of insect products, it is worthwhile noting that chitin affects this calculation due to its high non-protein nitrogen (N) content. Therefore, N should be multiplied by a factor of 5.60 for BSF protein and 4.76 for whole larvae rather than the usual 6.25 factor commonly used (Janssen et al., 2017; Ewald et al., 2020). As chitin content varies with insect life stage, it is also important to note that comparisons between sources may be affected differently. Evaluating insects on a true protein (sum of amino acids) basis over a crude protein basis is recommended to avoid these inaccuracies (Hawkey et al., 2021). This further allows better appreciation of soya replacement potential, the basis of which is digestible lysine, and not crude protein.

3   Production performance of pigs fed diets containing black soldier fly larvae

The existing literature demonstrates that BSF larvae are suitable to be included as a feed ingredient in pig diets (Newton et al., 1977; Spranghers et al., 2018; Biasato et al., 2019; Zhu et al., 2022; van Heugten et al., 2022). Research has shown that BSF's nutritional value, digestibility, and palatability (particularly larvae) are comparable to SBM and FM, highlighting its potential to partially replace these feed ingredients without adversely affecting performance and improving sustainability.

3.1 Growth performance of pigs feddiets containing protein from black soldier fly

In nursing piglets, partial replacement (3.5%) of FM in the creep diet with BSF larvae did not affect growth performance (Driemeyer, 2016) (Table 1).

This has also been observed in weaned piglets. Biasato et al. (2019) found that the inclusion of up to 10% partially defatted BSF larvae meal (replacing 60% of the SBM inclusion) in weaned piglet diets had no overall effect on growth performance. Average daily feed intake (ADFI) showed a linear response to increasing BSF larvae meal levels from 24 to 61 days post-weaning, with piglets fed 10% BSF larvae meal having the highest ADFI. This increase was attributed to increased palatability of the diet by the inclusion of partially defatted BSF. Spranghers et al. (2018) also reported no effect of full-fat (4% and 8% inclusion) and defatted (5.4% inclusion) BSF prepupae meal on the growth performance of weaned piglets when replacing whole toasted soya beans. However, the authors did note that, despite the lack of statistical differences between the treatment groups, piglets fed full-fat BSF prepupae meal at 4% or 8% inclusion gained less weight compared to control piglets fed toasted soya beans and piglets fed defatted BSF prepupae meal. Piglets fed full-fat BSF prepupae meal at 8% inclusion also had lower ADFI. They suggested this may have been caused by reduced palatability due to the large amount of free medium-chain fatty acids present in the feed containing full-fat BSF prepupae (Dierick et al., 2002; Spranghers et al., 2018). However, Newton et al. (1977) reported no differences in palatability between piglets fed BSF prepupae or SBM diets.

More recently, Boontiam et al. (2022) studied the effect of supplementing 6% or 12% full-fat BSF larvae on the growth performance of weaned piglets under poor hygiene conditions. Compared to the negative control (poor hygiene and 0% BSF larvae), pigs fed BSF larvae had higher body weights and ADFI from 1 to 28 days post-weaning. Dietary inclusion of BSF also improved body weight and ADFI from 15 to 28 days post-weaning compared to the positive control group (good hygiene and 0% BSF larvae). Over the entire experimental period (1 to 28 days post-weaning), no differences in growth

Table 1 Performance of pigs fed protein from black soldier fly (BSF).

Pig growth stage	Age at weaning	Experimental period	BSF life stage	BSF feed form	BSF inclusion (%)	Target replacement	Replacement (%)	Results	Reference
Nursing		10–28 d of age	Larvae	Meal	0, 3.5	Fishmeal	3.5	Growth performance is not affected. Haematological and biochemical parameters are not affected.	Driemeyer (2016)
Weaned piglets	20 d	1–61 d post-wean	Larvae	Partially defatted meal	0, 5, 10	Soybean meal	0, 31, 62	Growth performance is not affected. Haematological parameters are not affected. Gut morphology and histology are not affected. Trend for higher FI with 10% BSF inclusion.	Biasato et al. (2019)
Weaned piglets	28 d	1–28 d post-wean	Larvae	Full-fat meal	0, 6, 12	Soybean meal	0, 20, 40	Growth performance is not affected. Diarrhoea rate was reduced in pigs fed 6 or 12% BSF.	Boontiam et al. (2022)
	21 d	0–42 d post-wean	Larvae	Full-fat meal	0.5–15 (depending on phase)	Dried whey, fishmeal, blood meal, blood plasma	0, 25, 50	Growth performance is not affected.	Crosbie et al. (2021)
	32 d	0–27 d post-wean	Larvae	Meal	0, 5, 10, 20	Fishmeal	0, 25, 50, 100	Growth performance is not affected except for reduced ADG with 5% BSF inclusion.	Håkenåsen et al. (2021)
	28 d	1–28 d post-wean	Larvae	Meal	0, 4.5, 9.1, 13.7, 18.3	Soybean meal	0, 25, 50, 75, 100	ADG and ADFI were improved with 4.5 and 9.1% BSF inclusion. FCR was not affected.	Liu et al. (2023)

(Continued)

Table 1 (*Continued*)

Pig growth stage	Age at weaning	Experimental period	BSF life stage	BSF feed form	BSF inclusion (%)	Target replacement	Replacement (%)	Results	Reference
	21 d	1–15 d post-wean	Prepupae	Full-fat meal	0, 4, 8	Toasted soybean	0, 50, 100 72	Growth performance is not affected. Gut morphology and histology are not affected.	Spranghers et al. (2018)
	21 d	1–15 d post-wean	Prepupae	Defatted meal	5.4	Toasted soybean Fishmeal	0, 50, 100 72	Growth performance is not affected. Gut morphology and histology are not affected.	Spranghers et al. (2018)
	21 d	1–28 d post-wean	Larvae	Full-fat meal	0 1, 2, 4		0, 25, 50, 100	Growth performance is not affected. Serum total protein and globulin levels were highest while urea and triglyceride levels were lowest at 2% BSF inclusion. Jejunal villi height was the longest with 2% BSF inclusion. Inclusion of 2% BSF affected specific gut microbial populations, metabolic profiles, and mucosal immune gene expression.	(Yu et al. (2020a,b))

Pig growth stage	Age at weaning	Experimental period	BSF life stage	BSF feed form	BSF inclusion (%)	Target replacement	Replacement (%)	Results	Reference
Grower		9 wk	Larvae	Full-fat meal	0, 9, 12, 14.5, 18.5	Fishmeal	0, 25, 50, 75, 100	Growth performance is not affected. FCR was higher at 14.5 and 18.5% inclusion. Neutrophil count was higher and platelet counts were lower at 14.5 and 18.5% BSF inclusion. No other haematological parameters were affected.	Chia et al. (2019)
Grower		4 wk	Larvae		0, 3	Poultry offal	100	Growth and intake were not affected though FCR was increased.	Go et al. (2022)
Finisher		46 d	Larvae	Dried powder meal	0, 4, 8	Soybean meal	0, 18, 36	Final BW and ADG were the highest and FCR lowest at 4% inclusion. BSF inclusion positively impacted meat quality. BSF inclusion impacted microbial populations, metabolic profiles, and immune gene expression in the gut.	Yu et al. (2019a, 2019b)
Finisher		14 wk	Larvae	Full-fat meal	0, 6, 9, 12, 14	Fishmeal	0, 25, 50, 75, 100	ADFI was not affected; the Inclusion of 9 to 14% BSF improved ADG and reduced FCR.	Chia et al. (2021)

ADG = average daily gain; ADFI = average daily feed intake; BW = body weight; FCR = feed conversion ratio (feed over gain); FI = feed intake performance were found between the treatment groups; however, diarrhoea rate was significantly reduced in piglets fed BSF larvae (Boontiam et al., 2022). This suggests that feeding BSF can help improve piglet health and mitigate the effects of housing pigs in poor hygiene conditions (lighter body weight, increased diarrhoea, decreased nutrient utilisation). This is likely due to the antimicrobial properties of the fatty acid profile in the full-fat BSF, though prebiotic benefits arising from chitin cannot be excluded.

Furthermore, the inclusion of full-fat BSF larvae meal to partially replace (25% or 50%) animal protein sources (FM, spray-dried blood meal, and blood plasma) in nursery pig diets supported growth performance suggesting that dietary animal protein sources could be replaced by as much as 50% with full-fat BSF larvae meal (inclusion of up to approximately 15% in the diet) without negative consequences (Crosbie et al., 2021).

Partially replacing FM with 2% full-fat BSF larvae meal positively impacted ADG and FCR (F:G) in weaned piglets during the first two weeks post-weaning, with ADG being highest and FCR lowest in piglets fed 2% BSF larvae meal (Yu et al., 2020a). Over the 28-day post-weaning experimental period, no differences were found between the treatment groups fed 0% to 4% full-fat BSF larvae meal in replacement of 0% to 100% of the FM inclusion (Yu et al., 2020a). This is similar to what was reported by Håkenåsen et al. (2021) except the BSF inclusion rate was higher (0%, 5%, 10%, 20% of the diet), the BSF still replaced FM at 0%, 25%, 50%, 75%, and 100%. Interestingly, the only effect was seen for ADG over the 27-day post-weaning experimental period with piglets fed 5% full-fat BSF larvae meal having lower ADG compared to the control diet with the 10% and 20% BSF inclusion groups being intermediate (Håkenåsen et al., 2021). As this is the lowest BSF inclusion studied in this chapter, it may be that a higher inclusion and replacement of traditional protein sources are needed for the accurate evaluation of novel proteins in the diet.

Liu et al. (2023) investigated the effect of total replacement of SBM with increasing levels of BSF larvae meal (0% to 18.3%). The authors found that ADG and ADFI improved in piglets fed 4.5 and 9.1% BSF larvae meal (replacing 25% and 50% SBM, respectively), although FCR was unaffected.

In grower pigs, replacing up to 100% of fishmeal in the diet with full-fat BSF larvae meal (18.5% inclusion) did not negatively impact growth performance (Chia et al., 2019). Pigs fed 9% BSF larvae meal had better FCR than those fed 14.5 or 18.5% BSF larvae meal; however, all three groups were not different from the fishmeal control diet (Chia et al., 2019). Haematological parameters were generally unaffected, except for neutrophil and platelet counts, which were higher in the 14.5 and 18.5% BSF larvae meal inclusion groups (representing 75% and 100% fishmeal replacement, respectively).

Go et al. (2022) found that ADG was reduced during the first 2 weeks of feeding 3% BSF larvae compared to pigs fed a control diet containing poultry offal (100% replacement). However, by the end of the 4-week experimental period, no differences in ADG were noted. Pigs fed the 3% BSF larvae diet had lower FCR (G:F) than controls over the 4-week experiment. These minor differences in growth performance were attributed to anti-nutritional factors (ANFs) in the BSF, such as chitin. Chitin is a component of the insect's exoskeleton, and if not digested, it can bind to proteins and reduce digestibility (Wang and Shelomi, 2017).

In finisher pigs, partially replacing soybean meal with 4% dried BSF larvae meal (18% replacement) improved body weight and ADG and reduced FCR (F:G) compared to the control and 8% BSF inclusion diets (Yu et al., 2019a). This agrees with what has been reported in more recent studies. Chia et al. (2021) found replacing up to 100% fishmeal with up to 14% full-fat BSF larvae meal did not affect body weight or ADFI; however, the higher levels of BSF inclusion (9% to 14%) improved ADG and lowered FCR (F:G). Yu et al. (2019a) suggested that the differences between their treatment groups could be related to the upregulated expression of genes related to lipogenic potential and muscle fibre composition. They also attributed the lack of effect of the 8% BSF inclusion to higher chitin levels in the BSF diet. The improved growth performance in pigs fed increasing levels of BSF larvae meal in the study by Chia et al. (2021) was attributed to increased palatability.

The aforementioned studies were all carried out with the idea of using BSF as an alternative protein source. One study though looked at the use of BSF oil. Thus, when BSF larvae were included in the diet of weaned piglets as an extracted insect oil to replace corn oil, body weight, and ADG linearly increased with increasing BSF oil inclusion (0%, 2%, 4%, 6%). Serum cholesterol and platelet count also increased linearly with increasing BSF oil inclusion (van Heugten et al., 2022). This suggests that BSF larvae oil is at least as palatable as corn oil suitable to be considered as an alternative energy source.

Overall, these studies suggest that the inclusion of BSF in the diets of piglets both immediately post-weaning and at later ages does not adversely affect growth performance. The number of studies carried out in growing and finisher pigs is rather limited and so more work would be required to confirm the benefits of replacing SBM with BSF in these diets.

3.2 Nutrient digestibility

From the 12 studies reviewed, that reported pig performance when replacing traditional protein sources (one study included reviewed oil replacement), only three of these also reported nutrient digestibility. Commercial scale studies are still required to further evaluate the nutrient digestibility of BSF inclusion in pig diets, however we have summarised the current findings below.

Nutrient digestibility was not influenced by the inclusion of partially defatted 10% BSF larvae meal in the diet of weaned piglets (Biasato et al., 2019).

Dietary supplementation of 6% or 12% full-fat BSF larvae meal improved dry matter, crude protein, and ether extract digestibility in weaned piglets when compared to the negative control group (poor hygiene and 0% BSF); however, the BSF inclusion groups did not differ from the positive control (good hygiene and 0% BSF) (Boontiam et al., 2022). This improved digestibility is likely due to the improved gut health of piglets fed full-fat BSF larvae, which would help reduce unwanted inflammation and can result in increased energy partitioning to growth.

Including 5.4% defatted BSF larvae meal in weaned piglet diets resulted in equal or improved ileal nutrient digestibility compared to the control diet containing toasted soybean (Spranghers et al., 2018). Including full-fat BSF larvae meal at 4% or 8% reduced ileal energy digestibility; however, 4% BSF larvae meal inclusion increased ileal crude protein digestibility (Spranghers et al., 2018).

The dietary BSF inclusion in these studies is relatively low. At higher levels of BSF larvae inclusion (33%) in a study from almost 50 years ago, Newton et al. (1977) found reduced dry matter, increased ether extract, and similar crude protein digestibility in weaned piglets compared to those fed soybean meal. These differences may be due to changes in gut microbiome following the feeding of a novel feed ingredient.

3.3 Gut health, haematological parameters, and meat quality

Driemeyer (2016) found that including 3.5% BSF larvae meal in nursing piglet diets (replacing 100% fishmeal inclusion) did not affect haematological or biochemical parameters. The authors did observe increasing haemoglobin and haematocrit levels over the experimental period with BSF inclusion. Although the differences were not statistically significant, the authors suggested that they may be biologically important as higher haemoglobin and haematocrit levels can be indicators of immunological stress; however, the authors also noted that none of the animals showed physical signs of distress (Driemeyer, 2016).

Biasato et al. (2019) found that BSF inclusion in the diet of weaned piglets did not affect haematological parameters except for monocytes and neutrophils, which showed linear and quadratic responses to increasing BSF inclusion, respectively. Regardless, all of the haematological parameters recorded fell within the normal physiological range for pigs, suggesting that BSF inclusion did not negatively impact the piglet's health (Biasato et al., 2019). Inclusion of up to 10% partially defatted BSF larvae meal also did not influence gut morphology or histology (Biasato et al., 2019).

Serum cholesterol linearly increased in weaned piglets fed increasing levels of BSF larvae oil (van Heugten et al., 2022). Platelet count also tended to increase linearly with increasing BSF oil inclusion. BSF larvae oil inclusion affected no other haematological or serum parameters (van Heugten et al., 2022).

Yu et al. (2020a, 2020b) reported that weaned piglets fed 2% full-fat BSF larvae meal had higher serum total protein and globulin levels than piglets fed FM. Gut morphology was also influenced, with piglets fed 2% BSF larvae meal having longer jejunal villi heights than piglets fed fishmeal or 1% BSF larvae meal (Yu et al., 2020a). Including 2% BSF larvae meal also affected specific microbial populations, their metabolic profiles, and mucosal immune gene expressions in the gut (Yu et al., 2020b).

In grower pigs, neutrophil counts were increased, and platelet counts were reduced in pigs fed 14.5 or 18.5% full-fat BSF larvae meal instead of fishmeal (Chia et al., 2019), which coincided with greater (more detrimental) FCR. Since no other haematological were affected by BSF inclusion in this study, an elevated neutrophil level might be an indication of a stress response (Widowski et al, 1989), in accord with as suggested by Driemeyer (2016).

In finisher pigs, Yu et al. (2019a) found that including dried BSF larvae meal (4% or 8%) positively influenced carcass traits and meat quality as loin eye area and marbling scores were higher with BSF inclusion. Similar to what was found in the weaned piglets, Yu et al. (2019b) reported that BSF inclusion in finisher pig diets altered the microbial populations, their metabolic profiles, and mucosal immune gene expression in the gut.

4 Benefits of using black soldier fly

4.1 Reducing the use of traditional protein sources, soya bean meal, and fishmeal

This chapter evaluates peer-reviewed papers where BSF has been utilised to replace mainly SBM or FM in pig diets and these clearly indicate that while there is evidence to support the use of BSF as a suitable alternative, there is still a significant degree of variability regarding the upper limit and dose-dependency of this approach. For most studies, the largest inclusion of BSF used, ranging from 5% to 20%, did not impact growth performance. However, in some studies, positive effects on performance were observed at levels below the maximum tested. This supports that, as is the case with any ingredient, there are constraints to consider, which in the case of BSF are likely coming from the level of post-harvest processing regarding protein and oil extraction and chitin levels, which can be a result of variation in growth stage and processing. As such, owing to its positive effect on protein concentration, defatting BSF would be expected to increase its suitability in the commercial pig diet and therefore its soya replacement potential, though it should be noted that in the inclusion levels of full-fat, partially defatted, or defatted BSF reported to date, an upper limit on SBM replacement in young pigs seems not to have been demonstrated yet. There are also trade-offs to be considered between benefits and constraints from both BSF and the replaced ingredients. For the BSF, these trade-offs may come from the level of chitin, which at low inclusion level might be providing prebiotic benefits but over a threshold might result in fibre-based constraints. As such, replacing SBM with BSF will inevitably reduce the levels of fibre in the diet. Furthermore, if the replaced ingredient is arguably of greater digestibility, replacement on a digestible amino acid basis would be expected to detriment feed efficiency, as observed for BSF inclusion at the expense of poultry offal (Go et al., 2022). Finally, it should also be noted that whilst in some studies all the SBM was safely replaced, i.e. without impacting performance, this will be sensitive to the basal level of SBM in the first place. The lower the latter, the more likely it can be completely replaced with an alternative.

The vast majority of growth performance studies using BSF have been on peri-weaning pigs, whilst only a few have been carried out on grower and finisher pigs. Whilst arguably the level of SBM in pig rations reduces as pigs move through the different feeding phases, the greater feed intake of grower and particularly finisher pigs compared to peri-weaning pigs means that most SBM is being used in the grower and finisher phase. Only one study was identified in finisher pigs, where up to 8% of BSF meal was replacing up to 36% of SBM, with improved FCR and ADG at intermediate BSF levels (Yu et al., 2019). More studies in grower and finisher pigs are required to identify if a greater replacement of SBM for BSF results in similar performance, as this single study might suggest it may not be the case. The notion that alternative feedstuffs each have their own constraints when being used as 'soya replacers' comes largely from studies where these alternatives are addressed in isolation. However, an alternative approach would be to set a limit in the diet formulation to remove or restrain SBM inclusion and provide all other alternatives. A best-cost formulation may also look to include sustainability metrics to create a 'low carbon' or 'home-grown' diet which has other benefits to sustainable livestock production. Including BSF and other alternatives would help to result in a greater reduction in SBM, with the emphasis on using upper limits of each alternative below their anti-nutritional thresholds but so as to not produce a diet with too many ingredients. Such approaches have been shown successful in replacing all SBM in fast growing broilers (Houdijk et al., 2024), which arguably have a digestive system similar to that of pigs. Thus, realising that this can be achieved in broilers, this opens the notion to combine BSF with other protein sources such as pulses and oilseed meals from non-soya origin (e.g. rapeseed meal, sunflower meal). Indeed, within a background of a combination of SBM and rapeseed meal as main protein sources, a complete replacement of SBM with home-grown pulses for growing and finishing pigs has been observed under both experimental conditions (Smith et al., 2013; White et al., 2015) and commercial conditions (Houdijk et al., 2013).

Like SBM, FM has also been under pressure from a sustainability perspective, given that fish stocks are a finite resource (concerns of overfishing). FM also has environmental implications associated with its transport and processing (especially drying), and such factors have made it increasingly expensive and supply volatile. As such, we usually only find FM in the most specialist rations for young pigs (creep feeds, weaner feeds) as most of the FM available in the market is used in pet nutrition and aquaculture diets. The historic use of FM in young pig rations was not only driven by its high biological value, but also its associated health benefits, including gut health (see below) arising from a favourable fatty acid and amino acid profile. Therefore, replacing FM with BSF meal could not only provide a cost-effective alternative as well as one with improved sustainability credentials and supply benefits.

4.2 Gut health

Several studies have reported on the gut health benefits of BSF and extracted products such as insect oil, chitin, and antimicrobial peptides (AMPs) (Gasco et al., 2018; Biasato et al., 2019; Boontiam et al., 2022). While further work is needed to understand whether there is greater value in feeding these as extracted components or as the whole insect it should be noted that regulatory hurdles may also play a role in this decision.

As mentioned, chitin is the main carbohydrate incorporated in the insect exoskeleton and can impact protein and mineral bioavailability (Finke, 2007; Jonas-Levi and Martinez, 2017; Henriques et al., 2020). While pigs do not sufficiently produce chitinases to easily digest the chitin, it has been shown that the gut microbiota can secrete chitinase and gain value in this component (Šimůnek et al., 2001). The extent of this is likely related to the age of the pig and gut health before feeding a chitin-based compound.

Antimicrobial peptides are known to be produced by insects (Harlystiarini et al., 2019) as part of their immune response. These proteins inhibit the growth of harmful pathogens in the insect (Lu et al., 2014), but they also confer benefits to livestock feeding and may be a factor in the improvements in health seen in the papers mentioned above, most notably (Boontiam et al., 2022). *In vitro* BSF extracts have been shown to be strongly antibacterial against species of *Salmonella* and *E. coli* (Harlystiarini et al., 2019); both pathogens are important in pig production and may provide added value when feeding whole or extracted BSF products.

The composition of BSF oil is similar to that of coconut oil (Dayrit, 2015), predominantly containing lauric acid, with up to 52% recently having been reported (Ewald et al., 2020). Lauric acid is known to have antimicrobial properties (Spranghers et al., 2018). The content of lauric acid in BSF fat is affected by diet type and suitability but it is positively correlated to larvae weight and is hypothesised to be synthesised by the larvae (Spranghers et al., 2017) as it is found in high concentrations even when fed at very low levels in the insect substrate. Coconut oil and lauric acid are known to be beneficial in pig diets historically, with coconut oil being included in creep and nursery diets for antimicrobial benefits, but care must be taken to manage the antioxidant capacity of the diet when including highly saturated fatty acids.

Feeding of BSF has been shown to modify the gut microbiota (Håkenåsen et al., 2021; Boontiam et al., 2022; Liu et al., 2023), with increases seen in *Ruminococcaceae, Faecalibacterium, Butyricoccus,* and *Lactobacillus* spp. These changes may be due to the components mentioned above and changes to the macronutrients available in the diet. Further work is needed to consistently use BSF to improve gut health, but it is clear to see that gut health benefits can be gained from their feeding, and this should be harnessed to fully realise the value of BSF in pig diets.

4.3 Reducing carbon footprint of diets

It has been suggested that insect-based diets can reduce the carbon emissions of livestock feeding when replacing FM or SBM (van Huis and Oonincx, 2017; Ffoulkes et al., 2021). However, insect feedstocks and climate control in insect rearing must be considered (Oonincx, 2021) and the processing of insects post-harvest when compared to mass-produced feed protein such as SBM. In terms of locality and transport emissions, BSF can provide a highly localised protein source and may even be able to be produced on the site where the protein is to be used. This form of circular economy can make great use of locally available co-products, which would still fit with regulatory requirements within the EU. Insects such as BSF are classed as farmed livestock within the EU and as such must be reared and fed in accordance with relevant regulations (Alagappan et al., 2022). However, insects can be succesfully reared on vegetable surplus from supermarkets or pre-consumer sources. This would increase the nutrient suitability of these substrates and reduce unwanted variation in nutrient provision for livestock feeding (van Heugten et al., 2022). It may also be that insects play a role alongside anaerobic digesters, utilising waste heat and being used as a primary bio-converter in this process. We are increasingly seeing sustainable livestock farming look to utilise all on-farm primary and secondary resources to reduce waste, improve carbon capture and enhance overall resource efficiency; in this context, insects such as BSF can play a vital piece in this complicated puzzle (van Huis and Oonincx, 2017).

4.4 Animal welfare following feeding of BSF

A wealth of data supports the notion that insects support good animal welfare, especially regarding pig, poultry, and aquaculture feeding. This is mainly because most of the diet of juvenile monogastrics and fish would be supported by insects, either terrestrial or aquatic. Therefore, insect feeding supports key welfare criteria in livestock farming, providing towards freedoms 1 (freedom from thirst, hunger, and malnutrition) and 5 (freedom to express normal behaviour) (Mellor, 2016). The highest value regarding animal welfare from feeding insects may be attributed to live or whole insect feeding as they have then been attributed to providing stimulation in the form of manipulable materials in pig and poultry rearing (Star et al., 2020; Ipema et al., 2021a,b; 2022). However, wider use may be found in the use of insects in pelleted feeds, which would likely be in the form of a protein powder or extracted oil. This use may also benefit animal welfare, especially if the gut and animal health benefits reported above can be fully realised.

5 Challenges

5.1 Feed safety

A few studies have examined aspects of the safety of feeding BSF to pigs. Potential feed safety concerns such as microbial contamination and heavy metal accumulation need to be investigated and addressed (reviewed by van der Fels-Klerx et al. (2018)). An advantage is that, unlike other insect species, BSF are not considered disease vectors as the adults do not consume decayed organic material nor lay their eggs on organic material (van Huis, 2013). However, as with other animal-derived proteins, there is a risk of prion diseases, although there is (currently) no evidence that insects carry prions (DiGiacomo and Leury, 2019). BSF larvae have been found to accumulate some heavy metals such as cadmium from their diet but not others such as chromium, arsenic, nickel, and mercury (Charlton et al., 2015; Cai et al., 2018). Zinc concentration decreases in BSF larvae as its concentration increases in the rearing substrate (Diener et al., 2015). There are differing results regarding lead accumulation with some studies reporting higher concentrations in the BSF larvae than in the rearing substrate (Gao et al., 2017; Purschke et al., 2017; Cai et al., 2018) while others report the opposite (Diener et al., 2015). On a positive note regarding feed safety, BSF larvae have the ability to remove mycotoxins from contaminated feed without subsequent accumulation (Cai et al., 2018; van der Fels-Klerx et al., 2018). Additionally, BSF larvae reared on pesticide-spiked substrates did not show detectable levels of the pesticides in their tissues (Purschke et al., 2017). Thus, although when acting as bio-converters, some degree of bioaccumulation is always expected, the evidence that this is greatly impacting BSF quality is not strong. Therefore, existing quality assurance schemes in animal feed production that guarantee traditional feed safety are likely also applicable for the safe use of BSF.

5.2 Feed suitability

The knowledge of the nutritional value of BSF and any potential anti-nutritional factors it may contain will greatly determine its feed suitability. Some possible impacts of fibrous chitin have already been mentioned, including its prebiotic potential at relatively low levels and its fibre-like constraints at relatively high levels. The microbiome can make use of chitin as a source of energy by action of their chitinases. This opens the suggestions to develop chitinase as a possible feed additive, in the same way how, e.g. commercial phytases, xylanases, and proteases have been developed. This might assist the production of oligosaccharide-type structures from chitin to promote prebiotic properties and overall fermentability, and this contribution to host energy supply and physicochemical modification of gut content. Such an approach might also address the possible issue of allergen risk or allergy-promoting molecules present within the insects. Chitin may play a role in allergenicity as it is recognised by the immune system and can activate various immune cells (Komi et al., 2018). Despite chitin's role in immune and potentially allergic responses, the mechanisms by which it does so are still not fully understood (van Huis and Oonincx, 2017; van der Fels-Klerx et al., 2018). Another aspect of feed suitability is palatability. However, if palatability was a serious constraint, this would have been observed as reduction in feed intake in the performance studies undertaken. Since the latter has not been reported, it might be concluded that the palatability of BSF is unlikely a constraint.

6 Applications

6.1 Inclusion into pelleted feed

As has already been discussed earlier in the chapter, oil extracted BSF protein is a more suitable comparison to conventional protein sources (SBM and FM) due to the low level of oil in these materials. Pelleted monogastric feeds generally contain no more than 3.5% crude fat as higher levels can negatively affect pelleting and can reduce pellet quality as well as increasing the risk of oxidation and potentially leading to the early breakdown of feed vitamins.

Commercially, BSF is available and increasingly as a fat extracted protein over the whole unprocessed product. In terms of legislation, the EU classifies whole unprocessed terrestrial invertebrates and live-fed terrestrial invertebrates as currently permitted for feeding to all livestock. However, oil extraction would categorise the insect protein as processed animal protein (PAP) and is then regulated under different criteria. It is therefore in some cases more accessible to feed BSF as live and, or unprocessed rather than through the PAP regulations. For widespread use of BSF in pig diets it is likely that further processing will be needed to reduce limiting nutrients such as chitin and enable easy inclusion into pelleted feeds which makes up the majority of pig feed volume.

6.2 Live fed on farm

As noted above, this is generally an easier application due to the avoidance of the PAP regulations which came into force to reduce the risk of bovine spongiform encephalitis (BSE). Live feeding of BSF on farms is one of the main ways BSF are currently fed to livestock in the EU and UK, such as laying hens in the aim of reducing carbon footprint per egg. However, it is noted that current schemes are not fully effective due to the high costs of heating insect-rearing units and the limitations in legislation surrounding the feeding of insects. For the full value of BSF feeding to be realised it is likely that legislation surrounding permitted insect feeding substrates needs to be reviewed and relaxed to enable a wider array of lower value organic streams to be utilised.

7   Conclusion

The current body of research demonstrates that BSF has excellent potential as an insect species to be used for feeding commercial livestock, especially pigs. BSF is able to partially, and in some cases fully, replace SBM and FM without negative consequences for growth performance, gut health, or meat quality. BSF also has the ability to have less of an ecological footprint compared to the current plant protein sources but is hampered by current legislative frameworks. Ultimately, more research is needed at a commercial scale to ensure effective inclusion in commercial pig diets as a safe, cost-effective, and sustainable protein.

8   Acknowledgements

SRUC receives support from the Scottish Government (RESAS), including through an initiative around insect products for food production.

9 References

Alagappan, S., Rowland, D., Barwell, R., Mantilla, S. M. O., Mikkelsen, D., James, P., Yarger, O., and Hoffman, L. C. 2022. Legislative landscape of black soldier fly (Hermetia illucens) as feed. *J Insects Food Feed* 8:343–355.

Barragan-Fonseca, K. B., Dicke, M. and van Loon, J. J. A. 2017. Nutritional value of the black soldier fly (Hermetia illucens L.) and its suitability as animal feed: a review. *J Insects Food Feed* 3:105–120.

Biasato, I., Renna, M., Gai, F., Dabbou, S., Meneguz, M., Perona, G., Martinez, S., Lajusticia, A. C. B., Bergagna, S., Sardi, L., Capucchio, M. T., Bressan, E., Dama, A., Schiavone, A. and Gasco, L. 2019. Partially defatted black soldier fly larva meal inclusion in piglet diets: effects on the growth performance, nutrient digestibility, blood profile, gut morphology and histological features. *J Anim Sci Biotechnol* 10.

Boland, M. J., Rae, A. N., Vereijken, J. M., Meuwissen, M. P. M., Fischer, A. R. H., van Boekel, M. A. J. S., Rutherfurd, S. M., Gruppen, H., Moughan, P. J. and Hendriks, W. H. 2013. The future supply of animal-derived protein for human consumption. *Trends Food Sci Technol* 29:62–73.

Boontiam, W., Phaengphairee, P., Hong, J. and Kim, Y. Y. 2022. Full-fatted Hermetia illucens larva as a protein alternative: effects on weaning pig growth performance, gut health, and antioxidant status under poor sanitary conditions. *J Appl Anim Res* 50:732–739.

Cai, M., Hu, R., Zhang, K., Ma, S., Zheng, L., Yu, Z. and Zhang, J. 2018. Resistance of black soldier fly (Diptera: Stratiomyidae) larvae to combined heavy metals and potential application in municipal sewage sludge treatment. *Environ Sci Pollut Res* 25:1559–1567.

Charlton, A. J., Dickinson, M., Wakefield, M. E., Fitches, E., Kenis, M., Han, R., Zhu, F., Kone, N., Grant, N., Devic, E., Bruggeman, G., Prior, R. and Smith, R. 2015. Exploring the chemical safety of fly larvae as a source of protein for animal feed. *J Insects Food Feed* 1:7–16.

Chia, S. Y., Tanga, C. M., Osuga, I. M., Alaru, A. O., Mwangi, D. M., Githinji, M., Dubois, T., Ekesi, S., van Loon, J. J. A. and Dicke, M. 2021. Black soldier fly larval meal in feed enhances growth performance, carcass yield and meat quality of finishing pigs. *J Insects Food Feed* 7:433–447.

Chia, S. Y., Tanga, C. M., Osuga, I. M., Alaru, A. O., Mwangi, D. M., Githinji, M., Subramanian, S., Fiaboe, K. K. M., Ekesi, S., van Loon, J. J. A. and Dicke, M. 2019. Effect of dietary replacement of fishmeal by insect meal on growth performance, blood profiles and economics of growing pigs in Kenya. *Animals* 9. https://doi.org/10.3390/ani9100705

Crosbie, M., Zhu, C., Karrow, N. A. and Huber, L. A. 2021. The effects of partially replacing animal protein sources with full fat black soldier fly larvae meal (Hermetia illucens) in nursery diets on growth performance, gut morphology, and immune response of pigs. *Transl Anim Sci* 5. https://doi.org/10.1093/tas/txab057

Dayrit, F. M. 2015. The properties of lauric acid and their significance in coconut oil. *J Am Oil Chem Soc* 92:1–15.

Diener, S., Zurbrügg, C. and Tockner, K. 2015. Bioaccumulation of heavy metals in the black soldier fly, Hermetia illucens and effects on its life cycle. *J Insects Food Feed* 1:261–270.

Dierick, N. A., Decuypere, J. A., Molly, K., Van Beek, E. and Vanderbeke, E. 2002. The combined use of triacylglycerols containing medium-chain fatty acids (MCFAs) and exogenous lipolytic enzymes as an alternative for nutritional antibiotics in piglet nutrition. *Livest Prod Sci* 75:129–142.

DiGiacomo, K. and Leury, B. J. 2019. Review: insect meal: a future source of protein feed for pigs? *Animal* 13:3022–3030.

Driemeyer, H. 2016. Evaluation of black soldier fly (Hermetia illucens) larvae as an alternative protein source in pig creep diets in relation to production, blood and manure microbiology parameters. Available at https://scholar.sun.ac.za

Ewald, N., Vidakovic, A., Langeland, M., Kiessling, A., Sampels, S. and Lalander, C. 2020. Fatty acid composition of black soldier fly larvae (Hermetia illucens) – Possibilities and limitations for modification through diet. *Waste Manag* 102:40–47. https://doi .org/10.1016/j.wasman.2019.10.014

Ffoulkes, C., Illman, H., O'connor, R., Lemon, F., Behrendt, K., Wynn, S., Wright, P., Godber, O., Ramsden, M., Adams, J., Metcalfe, P., Walker, L.,. Gittins, J., Wickland, K., Nanua, S. and Sharples, B. 2021. Development of a roadmap to scale up insect protein production in the UK for use in animal feed. https://www.wwf.org.uk/sites/default/ files/2021-06/the_future_of_feed_technical_report.pdf

Finke, M. D. 2007. Estimate of chitin in raw whole insects. *Zoo Biol* 26:105–115.

Finke, M. D. 2013. Complete nutrient content of four species of feeder insects. *Zoo Biol* 32:27–36.

Gao, Q., Wang, X., Wang, W., Lei, C., and Zhu, F. 2017. Influences of chromium and cadmium on the development of black soldier fly larvae. *Environ Sci Pollut Res* 24:8637–8644.

Gasco, L., Finke, M. and van Huis, A. 2018. Can diets containing insects promote animal health? *J Insects Food Feed* 4:1–4.

Go, Y. B., Lee, J. H., Lee, B. K., Oh, H. J., Kim, Y. J., An, J. W., Chang, S. Y., Song, D. C., Cho, H. A., Park, H. R., Chun, J. Y. and Cho, J. H. 2022. Effect of insect protein and protease on growth performance, blood profiles, fecal microflora and gas emission in growing pig. *J Anim Sci Technol* 64:1036–1076.

Gupta, M., Hart, P., Krebsbach, S. G., Da Gama, L., Baungaard, C., Trewern, J., Halevy, S., Walsh, L., Blandon, A., Weir, C., Ryan, A., Wakefield, S., Tesco, D. E., Webb, L., Nordmann, H. D., Weis, B., Umeasiegbu, K., Ffoulkes, C., Illman, H., Behrendt, K., Godber, O., Ramsden, M., Adams, J., Metcalfe, P., Walker, L., Gittins, J., Wynn, S., O'connor, R., Lemon, F., Wickland, K., Nanua, S., Sharples, B., Wright, P., Keller, E., Perkins, R. and Salter, D. 2021. The future of feed: a WWF roadmap to accelerating insect protein in UK feeds. www.wwf.org.uk/press-release/insects-animal-feed -report. Text © 2021 WWF-UK. All rights reserved.

Håkenåsen, I. M., Grepperud, G. H., Hansen, J. Ø., Øverland, M., Ånestad, R. M. and Mydland, L. T. 2021. Full-fat insect meal in pelleted diets for weaned piglets: effects on growth performance, nutrient digestibility, gastrointestinal function, and microbiota. *Anim Feed Sci Technol* 281. https://doi.org/10.1016/j.anifeedsci.2021 .115086

Harlystiarini, H., Mutia, R., Wibawan, I. W. T. and Astuti, D. A. 2019. In vitro antibacterial activity of black soldier fly (Hermetia illucens) larva extracts against Gram-negative bacteria. *Buletin Peternakan* 43:125–129.

Hawkey, K., Brameld, J., Parr, T., Salter, A. and Hall, H. 2021. *Suitability of insects for animal feeding.Pages 26–38 in Insects as animal feed: novel ingredients for use in pet, aquaculture and livestock diets*. CABI, Wallingford, UK.

Hawkey, K. and Hall, H. 2023. Insects as animal feed. Pages 372–373 in *The Encyclopedia of Animal Nutrition*. Phillips, C., Ed. 2nd ed. CABI, GB.

Henchion, M., Hayes, M., Mullen, A. M., Fenelon, M. and Tiwari, B. 2017. Future protein supply and demand: strategies and factors influencing a sustainable equilibrium. *Foods* 6:1–21.

Henriques, B. S., Garcia, E. S., Azambuja, P. and Genta, F. A. 2020. Determination of chitin content in insects: an alternate method based on calcofluor staining. *Front Physiol* 11:1–10.

Heuel, M., Sandrock, C., Leiber, F., Mathys, A., Gold, M., Zurbrüegg, C., Gangnat, I. D. M., Kreuzer, M. and Terranova, M. 2022. Black soldier fly larvae meal and fat as a replacement for soybeans in organic broiler diets: effects on performance, body N retention, carcase and meat quality. *Br Poult Sci* 63:650–661.

Houdijk, J. G. M., Marchal, L., Bello, A., Gibbs, K. and Dersjant-Li, Y. 2024. Complete replacement of soya products with alternative ingredients for fast growing broilers. *British Poultry* Abstracts. in press.

Houdijk, J. G. M., Smith, L. A., Tarsitano, D., Tolkamp, B. J., Topp, C. E. F., Masey O'Neill, H. V., White, G., Wiseman, J., Kightley, S. and Kyriazakis, I. 2013. Peas and faba beans as home grown alternatives for soya bean meal in grower and finisher pig diets. Pages 145–175 in *Recent Advances in Animal Nutrition*. Garnsworthy, P.C., Wiseman, J., Eds. Nottingham University Press, Nottingham, UK.

Ipema, A. F., Bokkers, E. A. M., Gerrits, W. J. J., Kemp, B. and Bolhuis, J. E. 2021a. Providing live black soldier fly larvae (Hermetia illucens) improves welfare while maintaining performance of piglets post-weaning. *Sci Rep* 11. https://doi.org/10.1038/s41598 -021-86765-3

Ipema, A. F., Gerrits, W. J. J., Bokkers, E. A. M., Kemp, B. and Bolhuis, J. E. 2021b. Live black soldier fly larvae (Hermetia illucens) provisioning is a promising environmental enrichment for pigs as indicated by feed- and enrichment-preference tests. *Appl Anim Behav Sci* 244.

Ipema, A. F., Gerrits, W. J. J., Bokkers, E. A. M., van Marwijk, M. A., Laurenssen, B. F. A., Kemp, B. and Bolhuis, J. E. 2022. Assessing the effectiveness of providing live black soldier fly larvae (Hermetia illucens) to ease the weaning transition of piglets. *Front Vet Sci* 9.

Janssen, R. H., Vincken, J.-P., van den Broek, L. A. M., Fogliano, V. and Lakemond, C. M. M. 2017. Nitrogen-to-protein conversion factors for three edible insects: Tenebrio molitor, Alphitobius diaperinus, and Hermetia illucens. *J Agric Food Chem* 65:2275–2278.

Jonas-Levi, A. and Martinez, J. J. I. 2017. The high level of protein content reported in insects for food and feed is overestimated. *J Food Compos Anal* 62:184–188. http://dx.doi.org/10.1016/j.jfca.2017.06.004

Kim, S. W., Less, J. F., Wang, L., Yan, T., Kiron, V., Kaushik, S. J. and Lei, X. G. 2019. Meeting global feed protein demand: challenge, opportunity, and strategy. *Annu Rev Anim Biosci* 7:221–243.

Komi, D. E. A., Sharma, L. and Dela Cruz, C. S. 2018. Chitin and its effects on inflammatory and immune responses. *Clin Rev Allergy Immunol* 54:213–223.

Liu, S., Wang, J., Li, L., Duan, Y., Zhang, X., Wang, T., Zang, J., Piao, X., Ma, Y. and Li, D. 2023. Endogenous chitinase might lead to differences in growth performance and intestinal health of piglets fed different levels of black soldier fly larva meal. *Anim Nutr* 14:411–424. Available at https://linkinghub.elsevier.com/retrieve/pii/S2405654523000665

Lu, A., Zhang, Q., Zhang, J., Yang, B., Wu, K., Xie, W., Luan, Y. X. and Ling, E. 2014. Insect prophenoloxidase: the view beyond immunity. *Front Physiol* 5:1–15.

Mellor, D. J. 2016. Updating animalwelfare thinking: moving beyond the "five freedoms" towards "A lifeworth living." *Animals* 6. https://doi.org/10.3390/ani6030021

Newton, G. L., Booram, C. V., Barker, R. W. and Hale, O. M. 1977. Dried Hermetia illucens larvae meal as a supplement for swine. *J Anim Sci* 44:395–400. Available at https://academic.oup.com/jas/article/44/3/395-400/4698064

OECD/FAO. 2023. *OECD-FAO Agricultural Outlook 2023–2032*. OECD Publishing, Paris, FR.

Oonincx, D. G. A. B. 2021. *Environmental impact of insect rearing.Pages 53–59 in Insects as animal feed: novel ingredients for use in pet, aquaculture and livestock diets*. CABI, UK.

Purschke, B., Scheibelberger, R., Axmann, S., Adler, A. and Jäger, H. 2017. Impact of substrate contamination with mycotoxins, heavy metals and pesticides on the growth performance and composition of black soldier fly larvae (Hermetia illucens) for use in the feed and food value chain. *Food Addit Contam Part A Chem Anal Control Expo Risk Assess* 34:1410–1420. https://doi.org/10.1080/19440049.2017.1299946

Rathore, A. S. and Gupta, R. D. 2015. Chitinases from bacteria to human: properties, applications, and future perspectives. *Enzyme Res* 2015. https://doi.org/10.1155/2015/791907

Šimůnek, J., Hodrová, B., Bartoňová, C. and Kopečný, J. 2001. Chitinolytic bacteria of the mammal digestive tract. *Folia Microbiol (Praha)* 46:76–78.

Smith, L. A., Houdijk, J. G. M., Homer, D. and Kyriazakis, I. 2013. Effects of dietary inclusion of pea and faba bean as a replacement for soybean meal on grower and finisher pig performance and carcass quality. *J Anim Sci* 91:3733–3741.

Spranghers, T., Michiels, J., Vrancx, J., Ovyn, A., Eeckhout, M., De Clercq, P. and De Smet, S. 2018. Gut antimicrobial effects and nutritional value of black soldier fly (Hermetia illucens L.) prepupae for weaned piglets. *Anim Feed Sci Technol* 235:33–42.

Spranghers, T., Ottoboni, M., Klootwijk, C., Ovyn, A., Deboosere, S., De Meulenaer, B., Michiels, J., Eeckhout, M., De Clercq, P. and De Smet, S. 2017. Nutritional composition of black soldier fly (Hermetia illucens) prepupae reared on different organic waste substrates. *J Sci Food Agric* 97:2594–2600.

Star, L., Arsiwalla, T., Molist, F., Leushuis, R., Dalim, M. and Paul, A. 2020. Gradual provision of live black soldier fly (Hermetia illucens) larvae to older laying hens: effect on production performance, egg quality, feather condition and behavior. *Animals* 10. https://doi.org/10.3390/ani10020216

van der Fels-Klerx, H. J., Camenzuli, L., Belluco, S., Meijer, N. and Ricci, A. 2018. Food safety issues related to uses of insects for feeds and foods. *Compr Rev Food Sci Food Saf* 17:1172–1183.

van Heugten, E., Martinez, G., McComb, A. and Koutsos, E. A. 2022. Improvements in performance of nursery pigs provided with supplemental oil derived from black soldier fly (Hermetia illucens) larvae. *Animals* 12. https://doi.org/10.3390/ani12233251

van Huis, A. 2013. Potential of insects as food and feed in assuring food security. *Annu Rev Entomol* 58:563–583.

van Huis, A. and Oonincx, D. G. A. B. 2017. The environmental sustainability of insects as food and feed: a review. *Agron Sustain Dev* 37. https://doi.org/10.1007/s13593-017 -0452-8

Veldkamp, T. and Vernooij, A. G. 2021. Use of insect products in pig diets. *J Insects Food Feed* 7:781–793.

White, G. A., Smith, L. A., Houdijk, J. G. M., Homer, D., Kyriazakis, I. and Wiseman, J. 2015. Replacement of soya bean meal with peas and faba beans in growing/finishing pig diets: Effect on performance, carcass composition and nutrient excretion. *Anim Feed Sci Technol* 209:202–210.

Widowski, T. M., Curtis, S. E. and Graves, C. N. 1989. The neutrophil:lymphocyte ratio in pigs fed cortisol. *Can J Anim Sci* 69:501–504.

Yu, M., Li, Z., Chen, W., Rong, T., Wang, G., Li, J. and Ma, X. 2019a. Use of Hermetia illucens larvae as a dietary protein source: effects on growth performance, carcass traits, and meat quality in finishing pigs. *Meat Sci* 158:107837. https://doi.org/10.1016/j .meatsci.2019.05.008

Yu, M., Li, Z., Chen, W., Rong, T., Wang, G., Li, J. and Ma, X. 2019b. Hermetia illucens larvae as a potential dietary protein source altered the microbiota and modulated mucosal immune status in the colon of finishing pigs. *J Anim Sci Biotechnol* 10. https://doi .org/10.1186/s40104-019-0358-1

Yu, M., Li, Z., Chen, W., Rong, T., Wang, G., Li, J. and Ma, X. 2020a. Evaluation of full-fat Hermetia illucens larvae meal as a fishmeal replacement for weanling piglets: Effects on the growth performance, apparent nutrient digestibility, blood parameters and gut morphology. *Anim Feed Sci Technol* 264. https://doi.org/10.1016/j.anifeedsci .2020.114431

Yu, M., Li, Z., Chen, W., Rong, T., Wang, G., Li, J. and Ma, X. 2020b. Hermetia illucens larvae as a fishmeal replacement alters intestinal specific bacterial populations and immune homeostasis in weanling piglets. *J Anim Sci* 98:1–13.

Zhu, M., Liu, M., Yuan, B., Jin, X., Zhang, X., Xie, G., Wang, Z., Lv, Y., Wang, W. and Huang, Y. 2022. Growth performance and meat quality of growing pigs fed with black soldier fly (Hermetia illucens) larvae as alternative protein source. *Processes* 10. https://doi .org/10.3390/pr10081498

Chapter 3

The use of protein from yellow mealworms in fish feed

Enric Gisbert, Aquaculture Program – Institute of Agrifood Research and Technology (IRTA), Spain; and Alberto Ruiz, Aquaculture Program – Institute of Agrifood Research and Technology (IRTA), Spain and Aquaculture and Fisheries Group – Wageningen University and Research, The Netherlands

1 Introduction

Addressing the global challenge of feeding a growing population within environmental limits requires a shift towards circular food systems (van Riel et al., 2023). Amongst several food production methods, aquaculture has emerged as a crucial solution to meet rising food demands, alleviate poverty, enhance food security, and combat nutritional deficiencies, all with a lower environmental impact compared to alternative animal-source foods (Asche et al., 2022). Currently, aquaculture stands out as the world's fastest-growing food-producing sector, contributing a record 49.2% to the global production of

http://dx.doi.org/10.19103/AS.2024.0139.12

aquatic animals in 2020. The sector's remarkable expansion and intensification are evident across regions, with a total production of 122.6 million tonnes, valued at USD 264.8 billion. Inland waters contributed 54.4 million tonnes, whilst marine and coastal aquaculture yielded 68.1 million tonnes. Notably, most regions, except Africa, have witnessed continued aquaculture growth, driven by key producers like Chile, China, and Norway. However, challenges persist, hindering the sector's pursuit of sustainable outcomes. To ensure the sustainability and economic growth of aquaculture, a focus on innovative and sustainable practices, particularly in aquafeed formulation and feeding practices, is crucial (FAO, 2022).

Feed stands out as the principal input and a constraining element in aquaculture. Commercial aquafeeds are composed of blends of various feedstuffs designed to meet the fundamental nutritional needs of aquaculture species. Predominantly, marine-derived feed ingredients, notably fishmeal (FM) and fish oil (FO), serve as the primary sources of protein and lipids in formulated diets. Fishmeal is a dependable protein source due to its palatability, digestibility, and an optimal nutrient profile that satisfies the dietary requirements of aquatic species. Especially for carnivorous fish species, aquafeeds have heavily relied on FM as the primary protein source due to its superior essential amino acid (AA) profile, high protein content, enhanced nutrient digestibility, and the absence of antinutritional factors (Macusi et al., 2022). Marine-derived oils possess exceptional palatability, a highly digestible energy content, and a favourable nutritional profile rich in long-chain polyunsaturated fatty acids (PUFA), particularly arachidonic acid (ARA, 20:4n-6), eicosapentaenoic acid (EPA, 20:5n-3), and docosahexaenoic acid (DHA, 22:6n-3) (Mozanzadeh et al., 2022). Until 2005, FM and FO were consistently employed as cost-effective sources of protein and lipids in aquafeed formulations for both carnivorous and omnivorous species. Despite aquafeeds constituting only 4% of total industrial feed production (900–1000 Mt in 2018), the aquaculture sector consumes over 70% of global FM (Jannathulla et al., 2019).

However, the decline in fish catches and the persistent growth in FM consumption paint a bleak future for the aquaculture industry, unless there is a paradigm shift towards incorporating non-fish components in fish feed production (Macusi et al., 2022). Given the impracticality of relying on the volume of wild-caught forage fish required to sustain the surge in aquafeed production (projected at 87.1 million tons in 2025) based on existing aquafeed formulations, especially considering that capture fisheries have plateaued, and major fishing areas have reached their maximum potential (Naylor et al., 2021), it is imperative to reduce the dietary inclusion rates of FM and FO to ensure the continued growth of the aquaculture industry (Gephart et al,, 2020).

Reducing reliance on FM and FO in aquafeed is a crucial objective acknowledged by stakeholders as a top priority for the sustainable development of the aquafeed industry. The extent of dependence on FM and FO in feed formulations is largely contingent on the trophic position of the farmed species. Notably, for most omnivorous fish, dependency on FM and FO in diets is nearly negligible, and contemporary diets for carp, tilapia, or catfish are nearly devoid of FM and FO. However, decreasing FM and FO in aquafeeds poses a more significant challenge for carnivorous fish, especially marine species (Oliva-Teles et al., 2015). Over the past two decades, the aquafeed industry has made substantial strides in improving feed formulations and enhancing the efficiency of marine resource utilization for feed formulation. These advancements have been achieved through improvements in feed conversion ratios (FCRs), reductions in FM and FO inclusion levels in aquafeeds, and increased utilization of FM derived from trimmings and alternative raw ingredients (Naylor et al., 2021).

2   Finding alternatives to marine-derived aquafeed

Any alternative feedstuffs to traditional marine-derived ingredients, such as FM and FO, must fulfil a number of specific characteristics. These include nutritional suitability, ready availability, and ease of handling, shipping, storage, and use in feed production. Additionally, the chosen feeds should be evaluated based on their impacts on fish health and performance, consumer acceptance, minimal pollution and ecosystem stress, and potential human health benefits (Naylor et al., 2009). Furthermore, to foster the sustainable use of alternative and eco-friendly aquafeeds, there is a pressing need to opt for ingredients that can be supplied sustainably, grown locally and have a low environmental footprint (Mitra, 2021).

Considerable research has been dedicated to reformulating aquafeeds through the incorporation of novel ingredients and nutritional supplements, such as exogenous enzymes, bioactive compounds, and bioavailable trace metals. These additions complement plant proteins and address the specific needs of aquaculture species. A prominent challenge in the aquaculture industry is the creation of sustainable feeds that enhance fish welfare, maximize growth potential, and remain cost-efficient, especially when feeding costs can represent up to 50–60% of total production costs (Egerton, 2020).

Various nutritional strategies have been employed to replace marine-derived aquafeed ingredients like FM and FO, with approaches varying based on the targeted feedstuff. Plant-derived ingredients stand out as the most abundant and commonly-used alternative protein sources in aquafeeds (Hardy, 2010). Plant feedstuffs serve as major dietary protein sources for omnivorous and herbivorous fish and rank second to FM in diets for carnivorous species (Tacon, 2020). Moreover, plant proteins are frequently employed as replacements for FM in compound diets for aquatic species due to their cost-effectiveness and the negative consumer perception associated with the use of terrestrial animal by-products for fish feed (Egerton, 2020). However, plant feedstuffs come with challenges, including highly variable protein content, inadequate essential AA profiles and anti-nutritional factors, limiting their use in diet formulations (Oliva-Teles et al., 2015; Mitra, 2021). Recent concerns have also arisen about the social and environmental impact of soybean production for aquaculture feed which competes with growing the crop for human consumption or livestock feed (Tzachor, 2019). For these reasons, the aquafeed industry continues to explore new alternative feed protein ingredients to ensure sustainable growth.

Amongst the alternative feed ingredients, efforts have been focused on testing and validating protein sources like

- single-cell proteins (SCP);
- microalgae and macroalgae;
- rendered animal by-products; and
- insects.

Single-cell proteins, regardless of their origin (microalgae, yeast, fungi, or bacteria), offer crucial advantages over conventional protein ingredients, including a good nutritional profile, shorter production time, reduced land use, and independence from seasonal and climatic variations. Bacterial SCP, in particular, are considered popular and sustainable due to their high protein levels (up to 80% on a dry weight basis), AA profile resembling FM, and content in vitamins, phospholipids (PLs), and other functional compounds (Pereira, 2022).

Microalgae, recognized for their nutritional value and beneficial properties, have been acknowledged as an intriguing ingredient for aquafeeds, despite high production costs. However, this limitation is expected to be overcome in the near future (Aragão et al., 2022). Interest in macroalgae as alternative protein sources for FM is primarily due to their potential use as functional ingredients, including bioactive compounds and prebiotic effects (Camacho et al., 2019). Rendered animal by-products, such as poultry by-product meal, blood meal, and feather meal, contribute to the circular economy by offering nutrient profiles competitive in meeting the essential requirements of farmed fish species (Woodgate et al., 2022).

Finally, the last group of potential alternative proteins to be used in aquafeed that has received a lot of attention over the last year are insects. In particular, insects represent a sustainable ingredient with very interesting nutritional profiles and functional properties due to the bioactive compounds they contain (Nogales-Mérida et al., 2019; Gasco et al., 2020; Mousavi et al., 2020; Shafique et al., 2021). Although the use of insects in aquafeeds started only around 40 years ago, the industry has evolved rapidly with remarkable innovations in insect species selection, production systems, nutritional quality, testing, and validation. At least 16 insect species have already been evaluated as alternative protein sources in aquafeeds. Amongst the species tested and used for industrial aquafeed production, eight species are most promising:

1. Silkworms (*Bombyx mori*);
2. Black soldier fly (*Hermetia illucens*);
3. Housefly (*Musca domestica*);
4. Yellow mealworm;
5. Lesser mealworm (*Alphitobius diaperinus*);
6. House cricket (*Acheta domesticus*);
7. Banded cricket (*Gryllodes sigillatus*); and
8. Jamaican field cricket (*Gryllus assimilis*).

These eight species are the most studied species to date in replacing FM in aquafeeds and have also been approved for the production and inclusion in aquafeeds by the European Union (Alfiko et al., 2022).

3 Key advantages of yellow worm meal as an aquafeed

The effect of *Tenebrio molitor* meal (TMM) on fish growth and performance has been a subject of interest in aquaculture and fish nutrition research. The following features of TMM have driven interest in its use in feed for livestock and aquatic species:

1. High protein content: TMM is valued for its high protein content (Li et al., 2022; Tran et al., 2022b), which makes it a nutritious ingredient for fish feed. This is a crucial component in fish diets, essential for growth, muscle development, and overall animal health and condition (Aragão et al., 2022; Barroso et al., 2022).
2. Amino acid profile: the AA profile of TMM is generally well-balanced (Sánchez-Muros et al., 2016; Costa et al., 2020), providing essential AAs that are important for protein synthesis in fish. Amino acids play a key role in various physiological functions, including growth and immunity, and even in the sensory quality of the final product (Li et al., 2009).
3. Lipid content: TMM is also rich in fats (Costa et al., 2020; Hachero-Cruzado et al., 2024). Whilst moderate lipid content is beneficial for energy, excessive levels might need to be considered in diet formulations, especially for species with specific dietary requirements and essential fatty acid contents (Grigorakis, 2015).
4. Palatability: some studies have suggested that fish may find TMM very palatable, which can positively influence feed intake and growth

performance in fish, matching fish fed a FM-based diet (Ng et al., 2001) or even improving key performance indicators (KPIs) (Shafique et al., 2021).

5 Digestibility: the digestibility of TMM meal is an important factor affecting nutrient utilization by fish. High digestibility ensures that the nutrients are efficiently absorbed, contributing to growth and better feed utilization (Mastoraki et al., 2022; Chen et al., 2023).

6 Nutrient enrichment: TMM can serve as a source of various vitamins and minerals (Finke, 2022), contributing to the overall nutritional profile of aquafeeds (Costa et al., 2020; Shafique et al., 2021).

7 Sustainability: TMM is often considered a sustainable protein source (Magnani et al., 2023). The use of TMM in aquafeeds aligns with efforts to find alternative and eco-friendly protein sources for fish farming.

The purpose of this chapter is to provide an overview of the current use of *T. molitor* as alternative source of FM in aquafeeds. It summarizes recent reviews focused on the use of these invertebrates in aquafeeds and their impact on KPIs associated with growth performance and feed efficiency, product nutritional quality, and their immunomodulatory potential and effect on host gut microbiota.

4 Nutritional profile of yellow worm meal for aquafeed

In general terms, the composition of live yellow worm consists of *ca.* 20% protein, 13% fat, 2% fibre, and 62% moisture; in dried form, its composition may reach up approx 53% protein, 28% fat, 6% fibre, and 5% moisture (Mariod, 2020). When yellow worms are transformed into meal, their composition may vary depending on the developmental stage (Shafique et al., 2021) and production systems, rearing conditions, and feedstuffs used for their culture (Jajić et al., 2022) as summarized in Table 1. Figure 1 illustrates how the composition of selected TMM products varies compared to conventional animal-derived ingredients used in aquafeeds.

Looking at the AA profile of yellow worm, although it is generally well-balanced (Sánchez-Muros et al., 2016; Costa et al., 2020, Liland et al., 2021), TMM may have some AA imbalances when compared to FM, particularly in methionine, threonine, lysine, histidine, and cysteine, whilst TMM is rich in valine and tyrosine (Shafique et al., 2021). The AA profile of TMM shows higher lysine, methionine, and threonine content compared to other insect meals (Costa et al., 2020). The AA deficiencies from different TMM also depended on the fish species considered as shown in Figure 2.

For predatory fish species, chitin has the potential to be a significant carbohydrate source. However, chitin has proven to be difficult to digest by

Table 1 Proximate composition of yellow worm meals from different studies based on their development stage

Stage	CP	CF	Fibre	Ash	Moisture	References
Larvae	43.8	35.1	ND	2.9	7.8	Hachero-Cruzado et al. (2024)
Larvae	44.7	42.5	ND	3.7	2.4	Siemianowska et al. (2013)
Larvae	45.3	34.5	ND	4.1	ND	Costa et al. (2020)
D larvae	45.8	34.2	4.0	2.5	5.8	Hussain et al. (2017)
Larvae	46.4	32.7	4.6	2.9	5.3	Ravzanaadii et al. (2012)
D Larvae	47.0	29.6	ND	4.5	ND	Hoffmann et al. (2020)
Larvae	49.1	38.3	ND	4.1	ND	Liu et al. (2020)
Larvae	51.9	21.6	7.2	4.7	ND	Bovera et al. (2015)
FF Larvae	51.9	23.6	4.7	ND	ND	Gasco et al. (2016)
D Larvae	51.3	26.0	8.2	7.1	DD	Gelinçek and Yamaner (2020)
Larvae	51.9	23.6	6.6	4.7	5.1	Iaconisi et al. (2018)
Larvae	52.5	34.1	ND	3.5	3.6	Jeong et al. (2020)
Larvae	52.8	28.5	ND	2.9	6.5	Melenchón et al. (2023)
Larvae	52.8	36.1	ND	3.1	ND	Makkar et al. (2014)
D Larvae	53.0	3.6	3.1	26.8	ND	Khan et al. (2018)
Larvae	53.2	34.5	6.3	4.0	ND	Ghosh et al. (2017)
Larvae	54.9	23.6	6.6	4.4	5.3	Iaconisi et al. (2017)
Larvae	58.4	34.0	ND	4.8	3.5	Sánchez-Muros et al. (2016)
Larvae	58.4	30.1	3.5	8.0	ND	Barroso et al. (2014)
Larvae	60.2	19.1	22.4	4.2	7.3	Heidari-Parsa (2018)
Adult	63.3	7.6	20.0	3.6	3.5	Ravzanaadii et al. (2012)
Larvae	64.7	46.1	ND	5.9	11.6	Li et al. (2022)
Larvae	71.2	7.9	1.2	14.0	3.5	Tran et al. (2022b)

Note: References included in this table are representative of the myriad of studies including data on this issue. Columns fibre, ash, and moisture contents expressed as a percentage in dry weight basis. *Abbreviations:* CP, crude protein; CF, crude fat; D Larvae, dried larvae; FF Larvae, full fat larvae; ND, data not determined or provided by authors.

many fish species (Ringø et al., 2012), even though the inclusion of chitin in compound diets has been described as positive due to its immunomodulatory and prebiotic effects (see later discussion of the use of yellow worm meal as a functional ingredient). The contents of chitin in yellow worm larvae range between 2.9% and 10.1% depending on the stage of development and the methodology used for its quantification (Costa et al., 2020).

Fat and fatty acids constitute the second most abundant nutrient in the composition of yellow worm. As reviewed in Shafique et al. (2021), yellow worm larvae are rich in saturated (21.8–28.3% DM) and monounsaturated (38.1–52.5% DM) fatty acids, with palmitic acid (C16:0) and oleic acid (C18:1 n-9) amongst the most important ones. The content of PUFA varies from 22.7%

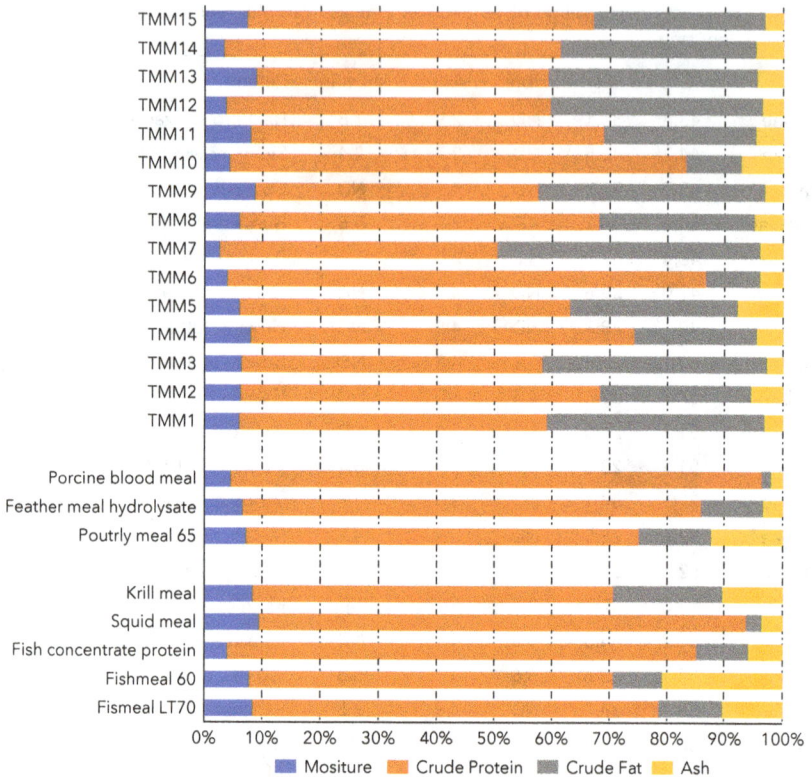

Figure 1 Comparison in terms of proximate composition (percentage in dry matter) of selected *Tenebrio molitor* meals (TMMs) with other marine and terrestrial animal protein sources used in aquafeed formulation. Selected yellow worm meals: TMM1, Gosh et al. (2017); TMM2, Khan et al. (2018); TMM3, Hussain et al. (2017); TMM4, Hoffmann et al. (2020); TMM5, Barroso et al. (2014); TMM6, Tran et al. (2022a); TMM7, Siemianowska et al. (2013); TMM8, Iaconisi et al. (2017); TMM9, Hachero-Cruzado et al. (2023); TMM10, Ravzanaadii et al. (2012); TMM11, Heidari-Parsa (2018); TMM12, Jeong et al. (2020); TMM13, Li et al. (2022); TMM14, Sánchez-Muros et al. (2016); TMN, 15, Melechón et al. (2023). Data from marine and terrestrial protein sources were retrieved from the Technical Booklet: a database of aquaculture feed ingredients (ARRAINA, 2013).

to 33.6%, with linoleic acid (C18:2 n-6) the most abundant one (21.8–31.9% DM). One disadvantage of yellow worm meals is their high lipid content and the lack of n-3 long-chain PUFA (LC-PUFA), such as EPA and DHA (Alfiko et al., 2022). Although defatting yellow worm meals is technically feasible, careful consideration is needed as the processes for fat extraction (i.e. use of organic solvents) and protein purification may reduce the profitability and environmental sustainability of this protein source and may affect its nutritional and functional value (Rawski et al., 2020).

	TMM1	Rainbow trout	Nile tilapia	Common carp	Channel catfish	TMM2	Rainbow trout	Nile tilapia	Common carp	Channel catfish	TMM3	Rainbow trout	Nile tilapia	Common carp	Channel catfish	TMM4	Rainbow trout	Nile tilapia	Common carp	Channel catfish	TMM5	Rainbow trout	Nile tilapia	Common carp	Channel catfish
Arginine	2,2	3,8	4,2	4,3	4,3	4,8	3,8	4,2	4,3	4,3	2,2	3,8	4,2	4,3	4,3	2,4	3,8	4,2	4,3	4,3	6,1	3,8	4,2	4,3	4,3
Histidine	2,8	1,1	1,7	2,1	1,5	3,4	1,1	1,7	2,1	1,5	1,7	1,1	1,7	2,1	1,5	1,4	1,1	1,7	2,1	1,5	3,6	1,1	1,7	2,1	1,5
Isoleucine	2,0	2,1	3,1	2,5	2,6	4,6	2,1	3,1	2,5	2,6	4,5	2,1	3,1	2,5	2,6	2,0	2,1	3,1	2,5	2,6	5,9	2,1	3,1	2,5	2,6
Leucine	3,4	2,6	3,4	3,3	3,5	8,6	2,6	3,4	3,3	3,5	5,3	2,6	3,4	3,3	3,5	3,6	2,6	3,4	3,3	3,5	8,7	2,6	3,4	3,3	3,5
Lysine	2,0	5,1	5,6	5,7	5,1	5,4	5,1	5,6	5,7	5,1	4,5	5,1	5,6	5,7	5,1	2,5	5,1	5,6	5,7	5,1	6,0	5,1	5,6	5,7	5,1
Threonine	1,8	2,6	3,8	3,9	2,2	4,0	2,6	3,8	3,9	2,2	1,6	2,6	3,8	3,9	2,2	1,9	2,6	3,8	3,9	2,2	4,5	2,6	3,8	3,9	2,2
Tryptophan	ND	0,4	1,0	0,8	0,5	0,6	0,4	1,0	0,8	0,5	ND	0,4	1,0	0,8	0,5	ND	0,4	1,0	0,8	0,5	ND	0,4	1,0	0,8	0,5
Valine	2,9	2,4	2,8	3,6	3,0	6,0	2,4	2,8	3,6	3,0	4,4	2,4	2,8	3,6	3,0	3,2	2,4	2,8	3,6	3,0	7,6	2,4	2,8	3,6	3,0
Methionine	ND	1,3	2,7	2,0	2,3	1,5	1,3	2,7	2,0	2,3	1,3	1,3	2,7	2,0	2,3	0,7	1,3	2,7	2,0	2,3	0,6	1,3	2,7	2,0	2,3
Cysteine	3,2	ND	ND	ND	ND	0,8	ND	ND	ND	ND	3,6	ND	ND	ND	ND	ND	ND	ND	ND	ND	ND	ND	ND	ND	ND
Phenylalanine	1,8	2,0	3,8	3,3	2,1	4,0	2,0	3,8	3,3	2,1	1,5	2,0	3,8	3,3	2,1	1,8	2,0	3,8	3,3	2,1	4,3	2,0	3,8	3,3	2,1
Tyrosine	3,5	ND	ND	ND	ND	7,4	ND	ND	ND	ND	2,3	ND	ND	ND	ND	3,0	ND	ND	ND	ND	4,2	ND	ND	ND	ND

Figure 2 Amino acid profile of several *Tenebrio molitor* meals (TMM) compared to the nutritional requirements of juveniles of rainbow trout (*Oncorhynchus mykiss*), Nile tilapia (*Oreochromis niloticus*), common carp (*Cyprinus carpio*) and channel catfish (*Ictalurus punctatus*). A semi-quantitative assessment indicated by different colours showed how close the AA profile of the TMM is from the particular requirement for each AA. For instance, dark green indicates that the AA composition of TMM was higher than 20% of its requirement; light green when the AA composition of the TMM was less than 20% above of its requirement; orange when the AA composition of TMM was deficient in less than 20% of its requirement, and red indicated that the AA composition was deficient in more than 20% of its requirements. Selected *T. molitor* meals: TMM1, Gosh et al. (2017); TMM2, Khan et al. (2018); TMM3, Hussain et al. (2017); TMM4, Hoffmann et al. (2020); TMM5, Barroso et al. (2014). Data on AA requirements for selected fish species has been retrieved from NRC (2011).

Few studies have evaluated the different lipid class composition of yellow worm larvae. Similar to other insects, triglycerides (TAGs) comprise 43.5–56.9% of total lipids. Looking at free fatty acids, percentages of diacylglycerides, monoacylglycerides, are 30.2–34.6%, 2.0–3.0%, and 1.3%, respectively (Costa et al., 2020; Hachero-Cruzado et al., 2024). As mentioned by these authors, TAGs are used for the storage of energy and fatty acids and they are the major lipid class for most insects, representing over 90% of total lipids. Phospholipid levels range from 4.1% to 5.7%, being the second most important lipid class. As PL are structural lipids of cell membranes, their content may considerably vary depending on the developmental stage considered. A wide range of cholesterol content has been described with value ranging from 0.4% to 4.2%, which has been attributed to different feedstuffs used for yellow worm rearing as cholesterol is absorbed from the diet or may result from the metabolism of dietary phytosterols (Costa et al., 2020; Hachero-Cruzado et al., 2024).

Yellow worm larvae are rich in minerals, with phosphorous (7.970 mg/kg) and potassium (8.007 mg/kg) the most abundant. Other minerals are present in smaller quantities such as magnesium (875–2.823 mg/kg), zinc (42–97 mg/kg), iron (38–68 mg/kg), copper (7.8–8.0 mg/kg), and manganese (4.4–11.6 mg/kg) (Costa et al., 2020; Shafique et al., 2021). However, it should be noted that yellow worm larvae may be deficient in calcium, vitamin D3, vitamin A, vitamin B12, thiamine, vitamin E, iodine, and sodium (Finke, 2002); however, this may be corrected by the use of mineral and vitamin premixes when feeds are manufactured.

5 Effects of yellow worm meal on growth and feed efficiency in different fish species

The effect of replacing FM with alternative protein sources not only depends on the target fish species (trophic position and feeding habits of the species), and the type of meal (i.e. full-fat insect meal, defatted or partially defatted insect meal) but also on insect production systems, stage of development, and processing technology used. The current section of this chapter focuses particularly on the evaluation and suitability of TMM as FM substitute, including its fat content, as well as its use in compound feeds for the main groups of farmed marine and freshwater fish species; in particular, salmonids, silurids (catfish), sparids, serranids, and other relevant aquaculture species.

5.1 Salmonids

Looking at salmonids, regardless of the economic importance of Atlantic salmon (*Salmo salar*) in the aquaculture market, most studies conducted in evaluating the feasibility of replacing FM with TMM have been conducted on rainbow trout (*Oncorhynchus mykiss*) and, to a lesser extent, on sea trout

(*Salmo trutta*). In contrast, studies dealing with insect meals in Atlantic salmon have been mainly focused on black soldier fly (*Hermetia illucens*) meals. The replacement of FM with TMM has been successfully tested in many nutritional studies in rainbow trout, even though results in terms of growth performance and composition vary, depending on the use of full-fat or defatted TMM.

The dietary replacement of FM with full-fat TMM up to 50% e.g. did not result in significant differences in growth performance and nutrient composition in rainbow trout (Gasco et al., 2014a). Gasco et al. (2016) found no significant difference in growth when 25% of FM was replaced with full-fat TMM; however, 50% replacement reduced fish growth rate without affecting flesh composition. The full replacement of FM (20%) with defatted TMM did not compromise growth and FCR values in rainbow trout (Chemello et al., 2020). Jeong et al. (2020) found that it was feasible to replace up to 28% of FM with TMM, but the best results in terms of growth performance and protein efficiency ratio (PER) were found at 14% FM replacement levels. Other studies testing the feasibility of replacing FM with defatted TMM showed that graded levels of FM replacement (20%, 30%, 60%, and 100%) with defatted TMM in rainbow trout did not affect body composition, whilst they increased protein and phosphorous retention (Rema et al., 2021). This is of special relevance since efficient phosphorus retention ensures that farmed fish convert the phosphorus in their feed into usable body mass, leading to better feed utilization and cost savings for farmers. In addition, improving phosphorus retention in fish reduces the environmental burden since phosphorus runoff from fish farms can lead to water pollution, causing eutrophication, harmful algal blooms, and oxygen depletion in aquatic ecosystems.

As Shafique et al. (2021) has discussed, levels of TMM required for FM replacement in aquafeeds depend on type of processing of yellow worm larvae. For instance, diets containing defatted TMM from larvae could effectively replace the entire content of FM. Other factors affecting TMM performance in aquafeeds may be associated with its chitin content, digestibility, AA balance, and fatty acid composition, which significantly impact the inclusion levels of TMM in aquafeeds, even though different diet formulation, ingredient selection, and experimental conditions may be also a source of variability when testing this alternative protein source.

5.2 Silurids (catfish)

Amongst other freshwater species, the suitability of TMM has been evaluated in catfish species. For instance, TMM could replace up to 75% of FM in diets containing up to 24% FM in yellow catfish (*Pelteobagrus fulvidraco*) (Su et al., 2017). The study also found that feed conversion efficiency in yellow catfish juveniles fed diets with up to 75% FM replacement by TMM showed similar results. Fishmeal can be replaced up to 40% by TMM without negative effects on growth performance and FCR values in African catfish (*Clarias gariepinus*), whilst fish fed diets with 20% TMM displayed better growth and feed efficiency indices than the control group (Ng et al., 2001). This study concluded that catfish fed diets with up to 80% replacement of FM with TMM still displayed good growth and feed utilization efficiency parameters. This is of special relevance since including TMM in compound diets does not compromise how feed consumed is converted into body mass. However, Roncarati et al. (2015) found a significant decrease in the body weight in common catfish (*Ameiurus melas*) fingerlings fed a diet with 50% of FM replaced by TMM compared to the control group, with no significant changes in the FCR amongst groups, whereas survival was slightly higher in the control diet (79% vs 70%, respectively).

5.3 Sparids

In gilthead seabream (*Sparus aurata*), replacing 25% of FM with TMM had no adverse effects on the growth, crude protein, and fat digestibility, whereas FCR and PER were improved compared to the control group. Considering that feed costs represent the largest operational expense in fish farming, often accounting for 50–70% of total production costs, improving feed efficiency means that fish convert more of the feed they consume into body mass, reducing the overall amount of feed needed. In contrast, at higher level of inclusion (50%), nutrient digestibility was reduced but this did not lead to negative effects on growth performance in comparison to the control group without TMM (Piccolo et al., 2017). Similarly, Fabrikov et al. (2020) demonstrated that FM could be replaced by up to 30% TMM without negative effects on growth performance, weight gain or FCR. Apparent digestibility coefficients (ADCs) of dry matter, organic matter, protein, and energy were not affected by 30% FM replacement by TMM, results that were comparable to those observed at similar replacement levels using other insect meals (Mastoraki et al., 2022). Regarding essential ADC values for essential AA, values (91.9%) were similar to those observed for premium quality FM (90.2%; Davies et al., 2009).

Research on testing TMM in sparids have not only focused on gilthead seabream, the most important species from this group of fishes from both a production and economic point of view. Ido et al. (2019) reported that the full FM replacement with defatted TMM in red seabream (*Pagrus major*) did not compromise growth and feed efficiency, and even promoted growth when compared to the control group fed a compound diet with 65% FM. Another study conducted in blackspot seabream (*Pagellus bogaraveo*) showed that inclusion of TMM larvae meal at 25% and 50% of FM substitution in the diet did not lead to significant effects on growth performance, feed intake or FCR (Iaconisi et al., 2017).

5.4 Serranids

European sea bass (*Dicentrarchus labrax*) is the most important marine fish species in the Mediterranean in terms of economic value and the second in terms of production. Gasco et al. (2014b) found that the dietary inclusion of 25% did not negatively impact somatic growth performance, whereas higher inclusion levels (50%) resulted in a reduction in growth, without compromising PER and feed intake. These results were partly in agreement with those reported by the same group of authors when diets were formulated with 50% full-fat TMM, where no differences were observed in FCR (Gasco et al., 2014b, 2016). Similarly, Reyes et al. (2020) did not find significant differences in FCR and PER, but did find a significant decrease in specific growth rate (SGR) and feed efficiency values in *D. labrax* fed a diet with 50% of FM replaced with TMM. Mastoraki et al. (2022) reported no significant differences in somatic growth and feed intake in *D. labrax* fed a diet with 30% of FM replacement by TMM, even though FCR were slightly worse than for the full-FM diet. Research on black sea bass (*Centropristis striata*), a commercially important species for US aquaculture, found it was possible to replace FM up to 50% with TMM without compromising growth performance, whereas TMM inclusion at 25% enhanced somatic growth without affecting FCR values when compared to the control group (Redman et al., 2019).

5.5 Flatfish and other species

Amongst flatfish species, Hachero-Cruzado et al. (2024) showed that it was feasible to replace up to 10% of FM with full-fat TMM in Senegalese sole (*Solea senegalensis*) with no negative effect on fish growth, FCR, nutrient utilization and survival. Replacing up to 15% of plant protein sources with TMM improved growth rate. In turbot (*Scophthalmus maximus*), FM can be replaced up to 30% TMM without affecting growth and feed efficiency, whereas higher levels of FM replacement (45-75%) resulted in a loss of weight and poor liver condition (Bai et al., 2023). *Tenebrio molitor* meal been used successfully as FM replacement up to 40% in olive flounder (*Paralichthys olivaceus*) without impairing growth, FCR and PER, whilst higher inclusion levels (60–80%) impaired growth and feed efficiency (Jeong et al., 2021).

Considering other carnivorous species, Chen et al. (2023) found that, in largemouth bass (*Micropterus salmoides*), it was possible to replace up to 36% of FM with TMM without affecting growth, whereas inclusion of 24% of TMM enhanced somatic growth compared to the control group. Based on a quadratic dose-response regression model, the authors recommended an optimal TMM inclusion of 19.5%. They determined that ADC of essential AAs in TMM was 93.9%. In perch (*Perca fluviatilis*), another perciform species, somatic growth,

FCR and PER were reduced when fishes were fed 50% and 75% TMM (Tran et al., 2022a). Formulated diets with up to 30% TMM caused no adverse effects on fish performance and feed utilization in mandarin fish (*Siniperca scherzeri*) juveniles, whereas lower inclusion levels (10% and 20%) led to better results in terms of growth and FCR values (Sankian et al., 2018). Regarding tilapiine species, Sánchez-Muros et al. (2016) described a decrease in somatic growth and PER, and an increase in FCR values when FM was replaced with TMM at 50% in Nile tilapia (*Oreochromis niloticus*). Amongst cyprinid species, TMM has been also tested in tench (*Tinca tinca*) (Rema et al., 2021). The replacement of FM at 18%, 28%, and 38% by TMM in diets negatively affected growth performance, regardless of the level of FM replacement considered. These results were also coupled with and increase in feed intake and lower FCR values. In mirror carp (*Cyprinus carpio* var. *specularis*), increasing TMM levels, ranging from 15% to 75%, enhanced somatic growth, reduced FCR and increased PER values without affecting feed intake (Li et al., 2022).

6   Effects of yellow worm meal on fillet quality

The market for farmed fish depends significantly on consumer perceptions of quality, with two crucial parameters at play: the nutritional and technical quality of fish. Nutritional quality relates to the food's nutrient content, specifically the presence of PUFA and their numerous health benefits. This not only contributes to the overall health benefits of fish but also shapes the consumer's perception of its nutritional value. On the other hand, technical quality is primarily an economic consideration, encompassing processing technical losses and edible yields, making it a pivotal factor for both consumers and processors (Grigorakis, 2015).

The relationship between diet and fillet composition and quality is complex and varies depending on the specific nutritional requirements of the fish species, the ingredient composition of fish diets and rearing conditions. The analysis of new ingredients and aquafeed formulations must not only consider their effect on KPIs related to growth and feed performance but also require considering fillet composition since it largely defines fish nutritional value and sensory quality perceived by the consumer.

The inclusion of higher levels than 25% of full-fat TMM has been found to result in reduced levels of n-3 EPA and DHA in fillets and higher levels of n-6 and n-9 fatty acids in rainbow trout (Iaconisi et al., 2018) and blackspot seabream (Iaconisi et al., 2017). The inclusion of TMM also affected the AA profile of the fillet as shown by Iacosini et al. (2017) who reported a reduction in essential AAs like alanine, isoleucine, leucine, and lysine contents in rainbow trout fillets when fed diets replacing 50% FM by full-fat TMM. However, Jeong et al. (2020) showed that the inclusion of 28% TMM in rainbow trout diets had no significant effects on whole body or fillet composition and AA profile compared to the control group (no TMM inclusion). Differences between studies might be due to factors such as age/size of the fish, composition, and nutrient content of the diet, source, or nutritional quality of the TMM, experimental conditions and water temperature, amongst other factors. As Belforti et al. (2015) have indicated, these variations amongst the trials make it difficult to compare the results from different studies and benchmarking studies using TTM from different production systems, stages of development or processing, need to be conducted under the same experimental conditions.

Regarding sparid species, Piccolo et al. (2017) found that the inclusion of 25% and 50% TMM in compound feeds for gilthead seabream did not affect its proximate body composition. Similar results were reported in blackspot seabream fed diets with 25% and 50% of FM replaced with full-fat TMM (Iaconisi et al., 2017). These authors also found an increase in linoleic acid levels with increasing dietary full-fat TMM levels. The same authors also found higher alanine, leucine, lysine, arginine, glycine, and proline levels in gilthead seabream fed diets with 50% of FM replaced with full-fat TMN larvae (Iaconisi et al., 2017). They also found lower levels of taurine, histidine, and phenylalanine in gilthead-seabream-fed diets with 25% and 50% of FM replaced by TMM. However, other authors have found that the dietary inclusion of TMM did not affect the levels of crude lipids in blackspot seabream fillets (Iaconisi et al., 2017).

Gasco et al. (2014b, 2016) found no differences in the overall body composition of European sea bass when fed with diets including 25% and 50% TMM. They also noted a reduction in DHA and EPA in fish given diets with 50% TMM (Gasco et al., 2014b, 2016). Replacement of FM by 80% TMM resulted in an increase in linoleic acid (C18:2) levels in European sea bass fillets but had no effects on EPA and DHA content (Basto et al., 2019). In largemouth bass, Chen et al. (2023) found that replacing FM with TMM at higher levels than 12% reduced body fat content and increased protein levels without impacting on ash and moisture levels.

Regarding flatfish species, Jeong et al. (2021) reported that increasing TMM levels at the expense of FM decreased the levels of crude fat in fillets without changing their protein content nor their essential AA composition in olive flounder. Furthermore, feeding TMM diets increased the levels of C18:1n-9 and C18:2n-6 in the fillet, at the expense of beneficial n-3 LC-PUFA. Concerning the effect of full-fat TM inclusion on the muscle and liver of Senegalese sole, it was observed that including full-fat insect meal in the diet led to a decrease in the total lipid content of the muscle, without changing the total lipid content of the liver (Hachero-Cruzado et al., 2024). Such reduction in muscle lipid content was associated to a decreases triacylglycerols and cholesterol. Lower lipid content in Senegalese sole fillets was thought to be due to the effect of chitin in TMM, which might have affected fat digestibility and lipid absorption.

Results from other species have shown that different levels of FM replacement by TMM did not modify the fillet composition in Nile tilapia (Sanchez-Muros et al., 2016), even though there was an increase in the levels of C18:1n-9 and C18:2n-6, as well as a decrease in n-3 LC-PUFA. In tench, replacement of FM by a combination of TMM with microalgae and macroalgae meals showed a linear increase in ash content with the increase of alternative protein sources to FM, changes that were associated with a decrease in crude lipid content when FM was replaced at 28% and 38% (Rema et al., 2021). In yellow croaker (*Larimichthys crocea*), levels of EPA, saturated fatty acids (SFAs), and the ratio of n-3/n-6 PUFA declined with increasing dietary replacement levels of FM by TMM, but mono-unsaturated fatty acids (MUFA), total PUFA, and n-6 PUFA increased with increasing dietary replacement levels of FM by TMM (Zhang et al., 2022). Overall, fine-tuning of the feed formulation may be required to avoid nutritional deficiencies in terms of essential AA and fatty acid profiles.

7 Effects of yellow worm meal on sensory and nutritional properties of fresh fish

The importance of fish feed on fillet characteristics such as colour, texture, and nutritional quality, as well as functional properties such as shelf life, is well-established. Changes in fish feed ingredients can affect the appearance, smell, and aroma of fish fillets, which may affect the perceived quality of the fillet and consumer acceptance (Magnani et al., 2023).

Amongst other parameters, skin colour plays a major role in the overall acceptability of fish, especially in those species that are sold as whole fish, since colour is commonly used as a key indicator of freshness and quality. In blackspot seabream, the inclusion of TMM did not affect the colour of the dorsal region of the skin, whereas the skin ventral region showed more differences (i.e. lower lightness and hue levels coupled with higher redness values in blackspot seabream fed 50% TMM) (Iaconisi et al., 2017). These changes in the skin pigmentation were associated with the β-carotene content (<200 μg/kg; Finke, 2002) in TMM, which may have enhanced the red colour of the skin (redness index). Amongst cyprinids, the combination of TMM with microalgae and macroalgae led to redder and yellower skin pigmentation in tench but lower lightness values, although the study did not separate out any specific effects of TMM on skin colour (Rema et al., 2021). Similar results regarding the enhancement of skin yellowness when TMM are included in compound diets have been reported in mirror carp and large yellow croaker (Li et al., 2022; Zhang et al., 2022). When evaluating the impact of TMM on skin pigmentation

Iaconisi et al. (2017) pointed out that changes in skin yellowness might be attributed to the riboflavin content in TMM (41–56 mg/kg; Finke, 2002) and the effects of this vitamin on skin chromatophores. The colour of fillets in blackspot seabream, yellowness, and intensity of the colour in the epaxial region of the fillet were also higher when fish were fed diets with TMM (Iaconisi et al., 2018). Similar results in terms of the effect of defatted and full TMM inclusion on fillet colour were reported by Melenchón et al. (2023) in rainbow trout.

Regarding the physical and textural properties of fillets, Iaconisi et al. (2017) found no significant differences in fillet quality parameters like water-holding capacity and texture characteristics (e.g. hardness, cohesiveness, resilience, gumminess, and adhesiveness). pH was lower in blackspot seabream fed with diets containing 50% TMM compared to the control diet (0% TMM) and 25% TMM. In contrast, the inclusion of different levels of TMM (15–75%) did not affect fillet pH values in mirror carp or muscle texture regardless of the level of FM replaced by TMM (Li et al., 2022). Similar results in terms of water-holding capacity and texture of flesh in large yellow croaker-fed diets replacing 15%, 30%, and 45% of FM by TMM were reported by Zhang et al. (2022), confirming the view that inclusion of TMM in aquafeeds does not alter textural properties, as also demonstrated in rainbow trout (Iaconisi et al., 2018).

In terms of consumer perception, Borgogno et al. (2017) reported differences in the perceived intensity of odour, flavour, and texture in rainbow trout fillets fed a diet that included insect meal from black soldier fly. However, the inclusion of TMM in aquafeeds did not affect sensory attributes and physicochemical parameters in fillets from rainbow trout when FM was replaced from 30% to 100% by TMM (Magnani et al., 2023), with similar results reported in blackspot seabream (Iaconisi et al., 2017). In rainbow trout, Magnani et al. (2023) did not find differences in flavour amongst experimental groups, even though numerical scores for earthy, sour, bitter, and for flavour intensity and persistence tended to increase with the full replacement of FM by TMM, which is in the line with other studies conducted with other insect meals (Borgogno et al., 2017). Magnani et al. (2023) found that insect meal-fed products were positively perceived by consumers as being nutritious, natural, and sustainable. However, their willingness to buy products was lower when compared to fish fed diets formulated with FM and plant protein sources. In contrast, Deely et al. (2022) found consumers were willing to pay more for environmentally friendly food products.

8 Effects of yellow worm meal on the host microbiome

The interplay between diet and microbiota in fish is an intricate and dynamic interaction pivotal for the overall health and well-being of aquatic organisms. The dietary choices of fish exert a significant influence on the composition and activity of their gut microbiota, with wide-ranging implications for host physiology, metabolism, immune function, and overall performance. Any disturbance in microbial composition or disruptions in host–microbe interactions, referred to as dysbiosis, can lead to digestive and systemic imbalances, potentially resulting in poor health and increasing disease risk (Sommer et al., 2017; Ruiz et al., 2024). These interactions are reciprocal, as the bacterial community can stimulate the immune system through pathogen-associated molecular patterns (PAMPs) recognized by pattern recognition receptors (PRRs), thereby activating immune signalling pathways. Concurrently, the host plays a role in shaping the microbial composition, regulating the abundance of potentially pathogenic bacterial communities (López-Nadal et al., 2020).

It is therefore important to assess how gut microbes respond to the inclusion of new raw materials replacing FM. One of the first studies evaluating the impact of TMM inclusion in aquafeeds was on rainbow trout where FM was replaced at 60% by TMM (Antonopoulou et al., 2019). They found that gut microbial diversity significantly decreased, indicating a more specialized gut bacterial community after FM replacement. However, Terova et al. (2021) showed no major effects of full FM substitution (100%) with TMM on microbial species richness and diversity of both gut mucosa- and skin mucus-associated bacterial communities in rainbow trout, results that were similar to those found in perch-fed diets containing 25–75% TMM (Tran et al., 2022a) or in sea trout (*Salmo trutta* m. *trutta*)-fed diets containing 40% of hydrolysed TMM (Mikołajczak et al., 2020; Józefiak et al., 2019). Ratios of Proteobacteria:Firmicutes, Proteobacteria:Bacteroidetes, and Firmicutes:Bacteroidetes indicated that replacement of FM at 60% did not affect the composition and functionality of gut microbiome in rainbow trout (Antonopoulou et al., 2019).

In European sea bass and gilthead seabream, Antonopoulou et al. (2019) reported that diversity of bacterial gut communities was not affected by the inclusion of TMM in diets, though effects varied by species. Fishmeal replacement by TMM resulted in an increased number of operational taxonomic units (OTUs), a proxy of bacterial species, in gilthead seabream and European sea bass, whereas the number of OTUs decreased in rainbow trout. Inclusion of TMM was thought to generate new nutritional niches in the intestines of both marine species, resulting in the appearance of new and unique OTUs when compared to the full FM-based diet. Changes in the relative abundance of the main phyla may result in potential gut imbalances that need to be further investigated by using omics approaches and physiological assays.

Józefiak et al. (2019) found that inclusion of 20% TMM in compound diets for rainbow trout resulted in a significant increase in the relative abundance of *Lactobacillus* and *Enterococcus* genera, which was attributed to the presence of chitin in the TMM since this compound has prebiotic properties. Similarly, an increase in *Lactobacillus* was found in the intestinal microbiota of perch-fed diets containing TMM (Tran et al., 2022a). This is of special relevance since the use of TMM may be regarded as beneficial for the host due to its potential probiotic properties (Ringø, 2020), even though its use as a functional ingredient needs to further explored and validated to take account of variables such as production system and feedstuff used.

9 The use of yellow worm meal as a functional ingredient

The effect of alternative feed ingredients or aquafeed formulations has to be addressed holistically, including their effects on host condition, immune and stress responses, and disease resistance against pathogens (Aragão et al., 2022). Insect meal should be regarded not only as a source of safe and high-quality protein for aquafeed but also as a potential source of dietary bioactive compounds. Research has focused on three categories of bioactive compounds found in insects:

- Antimicrobial peptides (AMPs);
- Fatty acids such as lauric acid; and
- Non-digestible polysaccharides like chitin and chitosan (Veldkamp et al., 2022).

Insect meal is recognized as being rich in AMPs, including defensins (effective against Gram-positive bacteria), diptericins, attacins, drosocins, and cecropins (active against Gram-negative bacteria), and drosomycins (which combat fungi) (Singh et al., 2013). These compounds exhibit antimicrobial activities through diverse mechanisms, including membrane pore formation or destruction of bacteria, interference with bacterial intracellular processes, or support for the host's immune system. The antioxidant capacity of insect proteins can also contribute to safeguarding the host against oxidative tissue damage (Veldkamp et al., 2022).

The inclusion of TMM in fish diets has been associated with increased antioxidant content in fish species like yellow catfish (Su et al., 2017), largemouth bass (Gu et al., 2022), mandarin fish (Sankian et al., 2018), and European sea bass (Henry et al., 2018). Antioxidants play roles in eliminating excessive superoxide radicals, contributing to the maintenance of reactive oxygen species (ROS) homeostasis in fish species (Sankian et al., 2018). However, other studies have shown that dietary inclusion of TMM did not promote activity of oxidative stress enzymes in rainbow trout (Jeong et al., 2020), pikeperch (Tran et al., 2022a), Nile tilapia (Sánchez-Muros et al., 2016), or tench (Hidalgo et al., 2022). These differences amongst studies might be due to differences in experimental conditions, species-specific nutritional requirements, and source and quality of the tested TMM. Henry et al. (2018) demonstrated a positive impact of TMM on the immune system in European sea bass. Full-fat TMM included at 24.8% exhibited anti-inflammatory and immunomodulating properties in the host, resulting in improvement of antibacterial and antiparasitic defences. In mandarin fish, an increase in lysozyme was found when fish were fed diets containing 30% full-fat TMM, though other immunological compounds remained unchanged (Sankian et al., 2018). In contrast, no differences in innate immune parameters were found in rainbow trout and tench-fed diets with graded levels of TMM (Valipour et al., 2019; Hidalgo et al., 2022). Although no differences in plasmatic innate immune parameters were found in tench-fed diets containing TMM, there was a trend for lower levels of lymphocytes, monocytes, and neutrophils, coupled with an increase in basophils, which was associated with a higher capacity for controlling bacterial infections, especially in those animals fed 5% TMM (13.8% FM replacement) (Hidalgo et al., 2022). However, other authors have found no differences in haematological indices in fish fed TMM diets (Sankian et al., 2018; Valipour et al., 2019; Mastoraki et al., 2022).

These differences between studies regarding the antioxidant and immunomodulatory properties of TMM may be associated with variables such as nutritional differences, but it is likely that enhancement of antioxidant activities in fish-fed diets containing TMM may be due to their chitin content (Ngo and Kim, 2014) as well as to other bioactive compounds present in yellow worms (Mousavi et al., 2020).

The immunomodulatory properties of feedstuffs and formulated diets may also be analysed by exposing the host to a particular pathogen and evaluating disease resistance mechanisms. Yellow catfish fed a diet containing 27% TMM (75% FM replacement) showed better survival rates than the control diet (0% TMM) when challenged with *Edwardsiella ictaluri* (Su et al., 2017). These results were coupled with an increase in lysozyme in serum and up-regulation of immune-related genes like immunoglobulin M (*igM*), major histocompatibility complex (*mhc*), and Interleukin-2 (*il-2*). This study suggests that dietary inclusion of TMM enhanced both innate and adaptive immune responses, allowing fish to better cope with bacterial infection. Similarly, red seabream fed with diets replacing 10% FM by defatted TMM and challenged with *Edwardsiella tarda* showed higher survival rates than those fed diets containing 0% or 5% defatted TMM (Ido et al., 2019). Similarly, pearl gentian grouper (♀ *Epinephelus fuscoguttatus* × ♂ *Epinephelus lanceolatu*) fed diets containing 7.5% TMM (18% FM replacement) and challenged with *Vibrio harveyi* showed higher survival rates than those specimens fed a diet devoid of TMM, whereas higher FM replacement (25% and 31%) compromised disease resistance in this species (Song et al., 2018).

10   Conclusion

The available literature shows that full-fat or defatted TMMs are a safe and sustainable alternative protein source for aquafeed formulations. The inclusion of TMM may reduce the fish-in and fish-out ratio, indicating that a lower quantity of marine forage fish is needed per unit of fish produced. This fact supports the sustainability of TMM with regard to the exploitation of fisheries stocks for marine raw materials for animal feeding. Furthermore, the use of yellow worm meal should not only be considered as a sustainable alternative to FM, but also for its functional characteristics. The immunomodulatory and antioxidant properties of TMM provides added value to this protein source, despite contradictory findings from different studies, which may be associated to its chitin composition and bioactive compound content.

Despite encouraging results in terms of growth and feed performance found in most of freshwater and marine species tested so far, further research needs to focus on yellow worm breeding and production technologies, including feedstuffs, as they have a direct effect on the composition and nutritional value of TMMs, as well as in their digestibility and chitin content (Barroso et al., 2022). Processing technologies used for producing insect meal need further investigation since each operation will have a different impact on the chemical and microbiological properties of the final product. This applies to novel food processing technologies (i.e. high pressure processing, pulsed electric field, ultrasound, and cold plasma) which have shown potential to modify, complement or replace conventional processing steps in insect processing (Ohja et al., 2021). The current heterogeneity in nutrient profiles of TMMs may limit its use in the aquafeed industry, as the feed manufacturer sector relies on the regular supply of ingredients with competitive prices and of homogenous and stable quality.

Although the inclusion of TMM in aquafeeds does not significantly change fillet composition, some changes in their fatty acid profile have been described. When including full-fat TMM in aquafeeds, particular attention should be paid to the final fatty acid profile of the feed, since yellow worm are rich in MUFA, PUFA, and n-6 PUFA and poor in n-3 PUFA, which indicates that FO or oil sources rich in n-3 PUFA may be needed to balance the fatty acid composition of the diet.

Finally, regardless of the advantages of using insect meal in aquafeeds, in order to foster its use for animal nutrition, more emphasis should be placed on the benefits of insect meal in animal feeds to achieve greater awareness and acceptance of the benefits of insect-based aquafeeds amongst consumers and aquaculture practitioners.

11   Where to look for further information

Further information about trends in research, investment, and innovation in insect-based feeds to meet aquafeed demands and maximize its environmental benefits may be found in the works of Fantatto et al. (2024) and of Selvaraj and Won (2024), whereas a detailed review on the effects of insect meals on fish immunity with special focus on mucosal tissues may be found in Islam et al. (2024).

Readers interested in understanding the role of insects in the circular bioeconomy are encouraged to consult Hamam et al. (2024). This study presents a systematic review of the literature, identifying the state-of-the-art and evaluating insects' contributions to closed-loop systems. It also explores opportunities and challenges associated with integrating insects into circular bioeconomy strategies, including applications in aquaculture.

The International Platform of Insects for Food and Feed, an EU non-profit representing the insect production sector to EU policymakers, stakeholders, and citizens, outlines its position on the use of insects in aquafeeds at this link: https://ipiff.org/.

Regarding the legislation involved in the use of insects in aquafeeds, the reader may find more information on the Regulation No. 2001/999 (Annex IV) amended by the Regulation 2017/893 (Annex X) that authorizes the use of insect proteins in aquafeeds from seven insect species: namely black soldier fly (*H. illucens*), common housefly (*M. domestica*), yellow mealworm (*T. molitor*), lesser mealworm (*A. diaperinus*), house cricket (*A. domesticus*), banded cricket (*G. sigillatus*), and field cricket (*G. assimilis*). Furthermore, the EU legislator authorized the use of silkworm (*B. mori*) processed animal proteins in aquaculture through the Regulation (EU) 2021/1925.

With the global aquaculture sector expected to grow in the coming years, research on alternative feed ingredients, such as those derived from insects, will continue to expand. This will lead to new studies, technical reports, and news focused on the use of insect meal and oil in finfish and crustacean nutrition and farming. For this reason, readers are encouraged to regularly look for newer and updated information in technical and scientific open access platforms like PubMed, arXiv, ResearchGate or Google Scholar, or in subscription-based platforms like the Web of Science™, ScienceDirect or SpringerLink among others.

12   References

Alfiko, Y., Xie, D., Astuti, R. T., et al. (2022), 'Insects as a feed ingredient for fish culture: status and trends', *Aquac. Fish.*, 7(2), 166–178.

ARRAINA (2013), Technical booklet. A database of aquaculture feed ingredients. Available at: https://www.sparos.pt/wp-content/uploads/2019/04/BOOKLET -ARRAIANA.pdf

Antonopoulou, E., Nikouli, E., Piccolo, G., et al. (2019). 'Reshaping gut bacterial communities after dietary Tenebrio molitor larvae meal supplementation in three fish species', *Aquaculture*, 503, 628–635.

Aragão, C., Gonçalves, A. T., Costas, B., et al. (2022), 'Alternative proteins for fish diets: implications beyond growth', *Animals*, 12(9), 1211.

Asche, F., Eggert, H., Oglend, A., et al. (2022), 'Aquaculture: externalities and policy options', *Rev. Environ. Econ. Policy*, 16(2), 282–305.

Bai, N., Li, Q., Pan, S. et al. (2023), 'Effects of defatted yellow mealworm (Tenebrio molitor) on growth performance, intestine, and liver health of turbot (*Scophthalmus maximus*)', *Anim. Feed Sci. Technol.*, 302, 115672.

Barroso, F. G., de Haro, C., Sánchez-Muros, M. J., et al. (2014), 'The potential of various insect species for use as food for fish', *Aquaculture*, 422, 193–201.

Barroso, F. G., Trnazado, C. E., Pérz-Jiménez, A., et al. (2022), 'Innovatve protein sources in aquafeeds', In J. M. Lorenzo and J. Simal-Gadara (Eds.), *Sustainable Aquafeeds, Technological Innovation and Novel Ingredients*, CRC Press, Boca Ratón, USA, pp. 139–184.

Basto, A., Maia, M.R., Pérez-Sánchez, J., et al. (2019), *Defatted yellow mealworm (Tenebrio molitor) larvae meal: a promising fishmeal substitute for European seabass*; Aquaculture Europe Conference (EAS), Berlin, Germany. Available at: https://digital .csic.es/handle/10261/202649

Belforti, M., Gai, F., Lussiana, C., et al. (2015). '*Tenebrio molitor* meal in rainbow trout (*Oncorhynchus mykiss*) diets: effects on animal performance, nutrient digestibility and chemical composition of fillets'. *Ital. J. Anim. Sci.*, 14(4), 4170.

Borgogno, M., Dinnella, C., Iaconisi, V., et al. (2017), 'Inclusion of *Hermetia illucens* larvae meal on rainbow trout (*Oncorhynchus mykiss*) feed: effect on sensory profile according to static and dynamic evaluations', *J. Sci. Food Agric.*, 97(10), 3402–3411.

Bovera, F., Piccolo, G., Gasco, L., et al. (2015), 'Yellow mealworm larvae (*Tenebrio molitor*, L.) as a possible alternative to soybean meal in broiler diets', *Brit. Poult. Sci.*, 56(5), 569–575.

Camacho, F., Macedo, A. and Malcata, F. (2019), 'Potential industrial applications and commercialization of microalgae in the functional food and feed industries: a short review', *Mar. Drugs*, 17(6), 312.

Costa, S., Pedro, S., Lourenço, H. et al. (2020), 'Evaluation of *Tenebrio molitor* larvae as an alternative food source', *NFS Journal*, 21, 57–64.

Chemello, G., Renna, M., Caimi, C., et al. (2020), 'Partially defatted Tenebrio molitor larva meal in diets for grow-out rainbow trout, *Oncorhynchus mykiss* (Walbaum): Effects on growth performance, diet digestibility and metabolic responses', *Animals*, 10(2), 229.

Chen, H., Yu, J., Ran, X., et al. (2023), 'Effects of yellow mealworm (*Tenebrio molitor*) on growth performance, hepatic health and digestibility in juvenile largemouth bass (*Micropterus salmoides*)', *Animals*, 13, 1389.

Davies, S. J., Gouveia, A., Laporte, J., et al. (2009), 'Nutrient digestibility profile of premium (category III grade) animal protein by-products for temperate marine fish species (European sea bass, gilthead sea bream and turbot)', *Aquac. Res.*, 40, 1759–1769.

Deely, J., Hynes, S., Barquín, J., et al. (2022), 'Are consumers willing to pay for beef that has been produced without the use of uncontrolled burning methods? A contingent valuation study in North-West Spain', *Econ. Anal. Policy*, 75, 577–590.

Egerton, S., Wan, A., Murphy, K., et al. (2020), 'Replacing fishmeal with plant protein in Atlantic salmon (*Salmo salar*) diets by supplementation with fish protein hydrolysate', *Sci. Rep.*, 10(1), 4194.

Fabrikov, D., Sánchez-Muros, M. J., Barroso, F. G., et al. (2020), 'Comparative study of growth performance and amino acid catabolism in *Oncorhynchus mykiss*, *Tinca tinca* and *Sparus aurata* and the catabolic changes in response to insect meal inclusion in the diet', *Aquaculture*, 529, 735731.

Fantatto, R.R., Mota, J., Ligeiro, C., et al. (2024). Exploring sustainable alternatives in aquaculture feeding: The role of insects. *Aquac. Rep.*, 37, 102228.

FAO (2022), *The State of World Fisheries and Aquaculture 2022. Towards Blue Transformation.* FAO, Rome. Available at: https://www.fao.org/3/cc0461en/cc0461en.pdf

Finke, M. D. (2002), 'Complete nutrient composition of commercially raised invertebrates used as food for insectivores', *Zoo. Biol.*, 21, 269–285.

Gasco, L., Belforti, M., Rotolo, L., et al. (2014a), 'Mealworm (*Tenebrio molitor*) as a potential ingredient in practical diets for rainbow trout (*Oncorhynchus mykiss*)', In *Insect to feed the world*, Wageningen UR, p. 69. Available at: https://flore.unifi.it/handle/2158/865861

Gasco, L., Gai, F., Piccolo, G., et al. (2014b), 'Substitution of fishmeal by *Tenebrio molitor* meal in the diet of *Dicentrarchus labrax* juveniles', In *Insect to feed the world*, Wageningen UR, p. 70. Available at: https://iris.unito.it/handle/2318/158357

Gasco, L., Henry, M., Piccolo, G., et al. (2016), '*Tenebrio molitor* meal in diets for European sea bass (*Dicentrarchus labrax* L.) juveniles: growth performance, whole body composition and in vivo apparent digestibility', *Anim. Feed Sci. Technol.*, 220, 34–45.

Gasco, L., Acuti, G., Bani, P., et al. (2020), 'Insect and fish by-products as sustainable alternatives to conventional animal proteins in animal nutrition', *It. J. Anim. Sci.*, 19, 360–372.

Gelinçek, İ. and Yamaner, G. (2020), 'An investigation on the gamete quality of Black Sea trout (*Salmo trutta labrax*) broodstock fed with mealworm (*Tenebrio molitor*)', *Aquac. Res.*, 51(6), 2379–2388.

Gephart, J. A., Golden, C. D., Asche, F., et al. (2020), 'Scenarios for global aquaculture and its role in human nutrition', *Rev. Fish. Sci. Aquac.*, 29(1), 122–138.

Grigorakis, K. (2015), 'Fillet proximate composition, lipid quality, yields, and organoleptic quality of Mediterranean-farmed marine fish: a review with emphasis on new species', *Crit. Rev. Food Sci. Nutr.*, 57(14), 2956–2969.

Ghosh, S., Lee, S. M., Jung, C., et al. (2017), 'Nutritional composition of five commercial edible insects in South Korea', *J. Asia-Pac. Entomol.*, 20(2), 686–694.

Gu, J., Liang, H., Ge, X., et al. (2022), 'A study of the potential effect of yellow mealworm (*Tenebrio molitor*) substitution for fish meal on growth, immune and antioxidant capacity in juvenile largemouth bass (*Micropterus salmoides*)', *Fish Shellfish Immunol.*, 120, 214–221.

Hamam, M., D'Amico, M., and Di Vita, G. (2024). Advances in the insect industry within a circular bioeconomy context: a research agenda. *Environ. Sci. Eur.*, 36(1), 29.

Hachero-Cruzado, I., Betancor, M. B., Coronel-Dominguez, A. J., et al. (2024), 'Assessment of full-fat *Tenebrio molitor* as feed ingredient for *Solea senegalensis*: effects on growth performance and lipid profile', *Animals*, 14(4), 595.

Hardy, R. W. (2010), 'Utilization of plant proteins in fish diets: effects of global demand and supplies of fishmeal', *Aquac. Res.*, 41(5), 770–776.

Heidari-Parsa, S. (2018), 'Determination of yellow mealworm (*Tenebrio molitor*) nutritional value as an animal and human food supplementation', *Arthropods*, 7(4), 94.

Henry, M. A., Gasco, L., Chatzifotis, S., et al. (2018), 'Does dietary insect meal affect the fish immune system? The case of mealworm, *Tenebrio molitor* on European sea bass, *Dicentrarchus labrax*', *Dev. Comp. Immunol.*, 81, 204–209.

Hidalgo, M. C., Morales, A. E., Pula, H. J., et al. (2022), Oxidative metabolism of gut and innate immune status in skin and blood of tench (*Tinca tinca*) fed with different insect meals (*Hermetia illucens* and *Tenebrio molitor*)', *Aquaculture*, 558, 738384.

Hoffmann, L., Rawski, M., Nogales-Merida, S., et al. (2020), 'Dietary inclusion of *Tenebrio molitor* meal in sea trout larvae rearing: Effects on fish growth performance, survival, condition, and GIT and liver enzymatic activity', *An. Anim. Sci.*, 20(2), 579–598.

Hussain, I., Khan, S., Sultan, A., et al. (2017), 'Meal worm (*Tenebrio molitor*) as potential alternative source of protein supplementation in broiler', *Int. J. Biosci.*, 10(4), 225–262.

Ido, A., Hashizume, A., Ohta, T., et al. (2019), 'Replacement of fish meal by defatted yellow mealworm (*Tenebrio molitor*) larvae in diet improves growth performance and disease resistance in red seabream (*Pargus major*)', *Animals*, 9(3), 100.

Iaconisi, V., Marono, S., Parisi, G., et al. (2017), 'Dietary inclusion of *Tenebrio molitor* larvae meal: effects on growth performance and final quality treats of blackspot sea bream (*Pagellus bogaraveo*)', *Aquaculture*, 476, 49–58.

Iaconisi, V., Bonelli, A., Pupino, R., et al. (2018), 'Mealworm as dietary protein source for rainbow trout: body and fillet quality traits', *Aquaculture*, 484, 197–204.

Islam, S. M., Siddik, M. A., Sørensen, M., et al. (2024). Insect meal in aquafeeds: a sustainable path to enhanced mucosal immunity in fish. *Fish Shellfish Immunol.* 150, 109625.

Jajić, I., Krstović, S., Petrović, M., et al. (2022), 'Changes in the chemical composition of the yellow mealworm (*Tenebrio molitor* L.) reared on different feedstuffs', *J. Anim. Feed Sci.*, 31(2), 191–200.

Jannathulla, R., Rajaram, V., Kalanjiam, R., et al. (2019), 'Fishmeal availability in the scenarios of climate change: inevitability of fishmeal replacement in aquafeeds and approaches for the utilization of plant protein sources', *Aquac. Res.*, 50(12), 3493–3506.

Jeong, S. M., Khosravi, S., Mauliasari, I. R., et al. (2020). Dietary inclusion of mealworm (*Tenebrio molitor*) meal as an alternative protein source in practical diets for rainbow trout (*Oncorhynchus mykiss*) fry', *Fish. Aquat. Sci.*, 23(1), 1–8.

Jeong, S. M., Khosravi, S., Yoon, K. Y., et al. (2021), 'Mealworm, *Tenebrio molitor*, as a feed ingredient for juvenile olive flounder, *Paralichthys olivaceus*', Aquac. Rep., 20, 100747.

Józefiak, A., Nogales-Mérida, S., Mikołajczak, Z., et al. (2019), 'The utilization of full-fat insect meal in rainbow trout (*Oncorhynchus mykiss*) nutrition: the effects on growth performance, intestinal microbiota and gastrointestinal tract histomorphology', *Ann. Anim. Sci.*, 19, 747–765.

Khan, S., Khan, R. U., Alam, W., et al. (2018), 'Evaluating the nutritive profile of three insect meals and their effects to replace soya bean in broiler diet', *J. Anim. Physiol. Anim. Nutr.*, 102(2), e662–e668.

Li, H., Hu, Z., Liu, S., et al. (2022), 'Influence of dietary soybean meal replacement with yellow mealworm (*Tenebrio molitor*) on growth performance, antioxidant capacity, skin color, and flesh quality of mirror carp (*Cyprinus carpio* var. *specularis*), *Aquaculture*, 561, 738686.

Li, P., Mai, K., Trushenski, J., et al. (2009), 'New developments in fish amino acid nutrition: towards functional and environmentally oriented aquafeeds', *Amino Acids*, 37, 43–53.

Liland, N. S., Araujo, P., Xu, X. X., et al. (2021), 'A meta-analysis on the nutritional value of insects in aquafeeds', *J. Insects Food Feed*, 7, 743–759.

Liu, C., Masri, J., Perez, V., et al. (2020), 'Growth performance and nutrient composition of mealworms (*Tenebrio molitor*) fed on fresh plant materials-supplemented diets', *Foods*, 9(2), 151.

Li, H., Hu, Z., Liu, S., et al. (2022), 'Influence of dietary soybean meal replacement with yellow mealworm (*Tenebrio molitor*) on growth performance, antioxidant capacity, skin color, and flesh quality of mirror carp (*Cyprinus carpio* var. *specularis*)', *Aquaculture*, 561, 738686.

López Nadal, A., Ikeda-Ohtsubo, W., Sipkema, D., et al. (2020), 'Feed, microbiota, and gut immunity: using the zebrafish model to understand fish health', *Front. Immunol.*, 11, 114.

Macusi, E. D., Cayacay, M. A., Borazon, E. Q., et al. (2022), 'Protein fishmeal replacement in aquaculture: a systematic review and implications on growth and adoption viability', *Sustainability*, 15(16), 12500.

Magnani, M., Claret, A., Gisbert, E., et al. (2023), 'Consumer expectation and perception of farmed rainbow trout (*Oncorhynchus mykiss*) fed with insect meal (*Tenebrio molitor*)', *Foods*, 12(23), 4356.

Makkar, H. P., Tran, G., Heuzé, V., et al. (2014), 'State-of-the-art on use of insects as animal feed', *Anim. Feed Sci. Technol.*, 197, 1–33.

Mariod, A. A. (2020), 'Nutrient composition of mealworm (*Tenebrio molitor*)', In A. A. Mariod (Ed.), *African edible insects as alternative source of food, oil, protein and bioactive components*, Springer Nature, Switzerland, pp. 275–280.

Mastoraki, M., Panteli, N., Kotzamanis, Y. P., et al. (2022), 'Nutrient digestibility of diets containing five different insect meals in gilthead sea bream (*Sparus aurata*) and European sea bass (*Dicentrarchus labrax*)', *Anim. Feed Sci. Technol.*, 292, 115425.

Melenchón, F., Larrán, A. M., Sanz, M. Á., et al. (2023), 'Different diets based on yellow mealworm (*Tenebrio molitor*)–part A: facing the decrease in omega– 3 fatty acids in fillets of rainbow trout (*Oncorhynchus mykiss*)', *Fishes*, 8(6), 286.

Mikołajczak, Z., Rawski, M., Mazurkiewicz, J., et al. (2020), 'The effect of hydrolyzed insect meals in sea trout fingerling (*Salmo trutta* m. *trutta*) diets on growth performance, microbiota and biochemical blood parameters', *Animals*, 10(6), 1031.

Mitra, A. (2021), 'Thought of alternate aquafeed: conundrum in aquaculture sustainability?', *Proc. Zoo. Soc.*, 74(1), 1–18.

Mousavi, S., Zahedinezhad, S., Loh, J. Y. (2020), 'A review on insect meals in aquaculture: the immunomodulatory and physiological effects', *Int. Aquat. Res.*, 12, 100–115.

Mozanzadeh, M. T., Hetmatpour, F., Gisbert, E. (2022), In J. M. Lorenzo and J. Simal-Gadara (Eds.), *Sustainable Aquafeeds, Technological Innovation and Novel Ingredients*, CRC Press, Boca Ratón, USA, pp. 185–292.

Naylor, R. L., Hardy, R. W., Bureau, D. P., et al. (2009), 'Feeding aquaculture in an era of finite resources', *Proc. Natl. Acad. Sci. USA*, 106(36), 15103–15110.

Naylor, R. L., Hardy, R. W., Buschmann, A. H., et al. (2021), 'A 20-year retrospective review of global aquaculture', *Nature*, 591(7851), 551–563.

Ng, W. K., Liew, F. L., Ang, L. P., et al. (2001), 'Potential of mealworm (*Tenebrio molitor*) as an alternative protein source in practical diets for African catfish, *Clarias gariepinus*', *Aquac. Res.*, 32, 273–280.

Ngo, D. H. and Kim, S. K. (2014), 'Antioxidant effects of chitin, chitosan, and their derivatives', *Adv. Food Nutr. Res.*, 73, 15–31.

Nogales-Mérida, S., Gobbi, P., Józefiak, D., et al. (2019), 'Insect meals in fish nutrition', *Rev. Aquac.*, 11, 1080–1103.

NRC (National Research Council) (2011), *Nutrient Requirements of Fish and Shrimp*, The National Academies Press, Washington, USA.

Ojha, S., Bußler, S., Psarianos, M., et al. (2021), 'Edible insect processing pathways and implementation of emerging technologies', *J. Insects Food Feed*, 7(5), 877–900.

Oliva-Teles, A., Enes, P., Peres, H. (2015), 'Replacing fishmeal and fish oil in industrial aquafeeds for carnivorous fish', In D. A. Davis (Ed.), *Feed and Feeding Practices in Aquaculture, Woodhead Publishing Series in Food Science, Technology and Nutrition*, Woodhead Publishing, Sawston, Waltham, pp. 203–233.

Pereira, A. G., Fraga-Corral, M., Garcia-Oliveira, P., et al. (2022), 'Single-cell proteins obtained by circular economy intended as a feed ingredient in aquaculture', *Foods*, 11(18), 2831.

Piccolo, G., Iaconisi, V., Marono, S., et al. (2017), 'Effect of *Tenebrio molitor* larvae meal on growth performance, in vivo nutrients digestibility, somatic and marketable indexes of gilthead sea bream (*Sparus aurata*)', *Anim. Feed Sci. Technol.*, 226, 12–20.

Ravzanaadii, N., Kim, S. H., Choi, W. H., et al. (2012), 'Nutritional value of mealworm, *Tenebrio molitor* as food source', *Int. J. Indus. Entomol.*, 25(1), 93–98.

Rawski, M., Mazurkiewicz, J., Kierończyk, B., et al. (2020), 'Black soldier fly full-fat larvae meal as an alternative to fish meal and fish oil in Siberian sturgeon nutrition: the effects on physical properties of the feed, animal growth performance, and feed acceptance and utilization', *Animals*, 10(11), 2119.

Redman, D. H., Nelson, D. A., Roy, J., et al. (2019), 'A pilot study using graded yellow mealworm (*Tenebrio molitor*) meal in formulated diets for growth performance of black sea bass (*Centropristis striata*)', *NOAA Technical Memorandum NMFS-NE-253*, 1–20.

Rema, P., Jorge, S., Leite, F., et al. (2021), 'Efficacy of microalgae (*Chlorella* and *Spirulina*) and insect (*Tenebrio molitor*) meals as fishmeal replacers in feed for juvenile tench (*Tinca tinca*)', *Revista Portuguesa de Zootecnia*, 6(1), 37–48.

Reyes, M., Rodríguez, M., Montes, J., et al. (2020), 'Nutritional and growth effect of insect meal inclusion on seabass (*Dicentrarchuss labrax*) feeds', *Fishes*, 5(2), 16.

Ringø, E., Zhou, Z., Olsen, R. E., et al. (2012), 'Use of chitin and krill in aquaculture–the effect on gut microbiota and the immune system: a review', *Aquac. Nutr.*, 18(2), 117–131.

Ringø, E. (2020), 'Probiotics in shellfish aquaculture', *Aquac. Fisher.*, 5(1), 1–27.

Roncarati, A., Gasco, L., Parisi, G., et al. (2015), 'Growth performance of common catfish (*Ameiurus melas* Raf.) fingerlings fed mealworm (*Tenebrio molitor*) diet', *J. Insects Food Feed*, 1(3), 233–240.

Ruiz, A., Gisbert, E., Andree, K. B. (2024), 'Impact of the diet in the gut microbiota after an inter-species microbial transplantation in fish', *Sci. Rep.*, 14(1), 4007.

Sánchez-Muros, M., De Haro, C., Sanz, A., et al. (2016), 'Nutritional evaluation of *Tenebrio molitor* meal as fishmeal substitute for tilapia (*Oreochromis niloticus*) diet', *Aquac. Nutr.*, 22(5), 943–955.

Sándor, Z. J., Banjac, V., Vidosavljević, S., et al. (2022), 'Apparent digestibility coefficients of black soldier fly (*Hermetia illucens*), yellow mealworm (*Tenebrio molitor*), and blue bottle fly (*Calliphora vicina*) insects for juvenile African catfish hybrids (*Clarias gariepinus* × *Heterobranchus longifilis*)', *Aquac. Nutr.*, 2022, 4717014.

Sankian, Z., Khosravi, S., Kim, Y. O., et al. (2018), Effects of dietary inclusion of yellow mealworm (*Tenebrio molitor*) meal on growth performance, feed utilization, body composition, plasma biochemical indices, selected immune parameters and antioxidant enzyme activities of mandarin fish (*Siniperca scherzeri*) juveniles', *Aquaculture*, 496, 79–87.

Selvaraj, V., Won, E. (2024). Transforming aquaculture with insect-based feed: restraining factors, *Animal Frontiers*, 14(4), 24–7.

Shafique, L., Abdel-Latif, H. M. R., Hassan, F. U., et al. (2021), 'The feasibility of using yellow mealworms (*Tenebrio molitor*): towards a sustainable aquafeed industry', *Animals*, 11, 811.

Siemianowska, E., Kosewska, A., Aljewicz, M., et al. (2013), Larvae of mealworm (*Tenebrio molitor* L.) as European novel food', *Agric. Sci.*, 4(6), 336935.

Singh, A., Phougat, N., Kumar, M., et al. (2013), 'Antifungal proteins: potent candidate for inhibition of pathogenic fungi', *Curr. Bioact. Compd.*, 9(2), 101–112.

Sommer, F., Anderson, J. M., Bharti, R., Raes, et al. (2017), 'The resilience of the intestinal microbiota influences health and disease'. *Nat. Rev. Microbiol.*, 15(10), 630-8.

Song, S. G., Chi, S. Y., Tan, B. P., et al. (2018), 'Effects of fishmeal replacement by *Tenebrio molitor* meal on growth performance, antioxidant enzyme activities and disease resistance of the juvenile pearl gentian grouper (*Epinephelus lanceolatus* ♂ × *Epinephelus fuscoguttatus* ♀)', *Aquac. Res.*, 49(6), 2210–2217.

Su, J., Gong, Y., Cao, S., et al. (2017), 'Effects of dietary *Tenebrio molitor* meal on the growth performance, immune response and disease resistance of yellow catfish (*Pelteobagrus fulvidraco*)', *Fish Shellfish Immunol.*, 69, 59–66.

Tacon, A. G. J. (2020), 'Trends in global aquaculture and aquafeed production: 2000–2017', *Rev. Fish. Sci. Aquacult.*, 28, 43–56.

Terova, G., Gini, E., Gasco, L. et al. (2021), 'Effects of full replacement of dietary fishmeal with insect meal from Tenebrio molitor on rainbow trout gut and skin microbiota', *J. Animal Sci. Biotechnol.*, 12, 30.

Tran, H. Q., Nguyen, T. T.2, Prokešová, M., et al. (2022a), 'Systematic review and meta-analysis of production performance of aquaculture species fed dietary insect meals', *Rev. Aquac.*, 14(3), 1637–1655.

Tran, H. Q., Prokešová, M., Zare, M., et al. (2022b), 'Production performance, nutrient digestibility, serum biochemistry, fillet composition, intestinal microbiota and environmental impacts of European perch (*Perca fluviatilis*) fed defatted mealworm (*Tenebrio molitor*)', *Aquaculture*, 547, 737499.

Tzachor, A. (2019), 'The future of feed: integrating technologies to decouple feed production from environmental impacts', *Ind. Biotechnol.*, 15(2), 52–62.

van Riel, A. J., Nederlof, M. A., Chary, K., et al. (2023), 'Feed-food competition in global aquaculture: current trends and prospects', *Rev. Aquac,* 15, 1142–1158.

Valipour, M., Oujifard, A., Hosseini, A., et al. (2019), 'Effects of dietary replacement of fishmeal by yellow mealworm (*Tenebrio molitor*) larvae meal on growth performance, hematological indices and some of non-specific immune responses of juvenile rainbow trout (*Oncorhynchus mykiss*)', *Iran. Sci. Fish. J.*, 28, 13–26.

Veldkamp, T., Dong, L., Paul, A., et al. (2022), 'Bioactive properties of insect products for monogastric animals–a review', *J. Insects Food Feed*, 8(9), 1027–1040.

Woodgate, S. L., Wan, A. H., Hartnett, F., et al. (2022), 'The utilisation of European processed animal proteins as safe, sustainable and circular ingredients for global aquafeeds', *Rev. Aquac.*, 14(3), 1572–1596.

Zhang, Z., Yuan, J., Tian, S., et al. (2022), 'Effects of dietary vitamin E supplementation on growth, feed utilization and flesh quality of large yellow croaker *Larimichthys crocea* fed with different levels of dietary yellow mealworm *Tenebrio molitor* meal', *Aquaculture*, 551, 737954.

Chapter 4

Emerging protein sources for poultry feed

Archibold G. Bakare, Fiji National University, Fiji Islands; Taiye Olugbemi, Ahmadu Bello University, Nigeria; Mohammed M. Ari, Nasarawa State University, Nigeria; and Paul A. Iji, Fiji National University, Fiji Islands and University of New England, Australia

1 Introduction

Commercial poultry diets contain from 15% to 24% total protein, depending on the age of birds, level of production and end product. Most of the protein is supplied from major plant and animal sources. Although the range of ingredients that can supply this protein is wide, the global poultry industry tends to depend on a limited number of feed ingredients due to supply, ingredient quality and safety issues.

The supply of key protein sources around the world is variable and this has a major impact on the costs of diets and production. The number one vegetable protein source is soybean, with supply coming mostly from North and South America. Other key vegetable sources such as peanut, canola, cotton seed and sunflower are produced across many more regions than soybean, but they tend to have more constraints in supply and quality. The sources of animal protein ingredients and their quality are even more variable. However,

http://dx.doi.org/10.19103/AS.2024.0143.19

such products, including meat and bone meal, and fishmeal remain the most important ingredients that can complement vegetable protein sources in most commercial diets. Globally, there are emerging sources of protein for the poultry industry. Many ingredients have always been available but have been under-explored or not considered as protein sources for poultry. The objective of this chapter is to examine some of these emerging ingredients for potential use locally or for the development into global protein sources for the poultry industry.

2 Challenges in using conventional protein sources in poultry diets

Conventional protein sources are of both plant and animal origins. Plant sources include soybean, peanut, cottonseed, sunflower seed, rapeseed and linseed, while animal sources include fishmeal, meat and bone meal. Soybean meal and fishmeal, however, are the most used conventional protein sources used in poultry feeds. Despite the many positive attributes of these conventional protein sources in poultry diets, their use is not without challenges associated with nutritional limitations, animal and human health concerns, availability and cost as well as environmental and sustainability issues (Sajid et al., 2023). Some sources are known to contain anti-nutritional factors that can reduce nutrient absorption and utilization by poultry, leading to poor growth and development. There is concern that the use of animal by-products in poultry feed can have an adverse impact on animal and human health, leading to food safety concerns and potential regulatory issues (Jedrejek et al., 2016; an overview e.g. of recent developments in EU policy and regulation in this area can be found at: https://www.efsa.europa.eu/en/topics/animal-by -products). Allergens contained in some conventional protein sources can cause allergic reactions in poultry and may lead to decreased performance or health of birds.

The availability of conventional protein sources varies according to location and season, which can lead to disruptions and fluctuations in supply, resulting in price fluctuations as well as availability. Some sources may not be available in certain countries or regions, leading to increased costs and decreased profitability. Growing demand for these ingredients has resulted in higher feed costs. The use of conventional protein sources for poultry feed may also not be sustainable in the long term, as it can lead to over-exploitation of natural resources (e.g. deforestation to grow protein crops) and environmental degradation (Fraanje and Garnett, 2020). The production of soybeans, for instance, also requires considerable resources and energy. Fishmeal can also have a significant impact on the marine ecosystem.

3 Potential alternative sources of protein: animal sources

As the world seeks sustainable protein sources, alternative protein sources like insect meal, earthworms, rumen digesta, tankage and blood meal present viable options in non-ruminant diets, especially poultry feeds. These alternatives contribute to the evolution of modern animal nutrition given their nutritional profiles, which, with proper processing, formulation and inclusion strategies, would be beneficial for the growth and well-being of poultry.

3.1 Insect meal

Insect meals are obtained by processing various insect species, such as mealworms, black soldier fly larvae and crickets. The nutrient diversity of insect meals effectively makes them suitable as a replacement for conventional protein sources like fish and meat meals. Insect meal protein content ranges from 39% to 65% (Moyo and Moyo, 2022). Amino acid profiles vary among insect species but often include essential amino acids. They also contain vitamins, minerals and useful fats (Kourimska and Adamkova, 2016; Spranghers et al., 2017). Sustainable sourcing of insects can be guaranteed since they can be reared on organic waste and reducing environmental impact. The high protein content of insect meals makes it highly suitable for poultry. With proper amino acid supplementation, insect meal inclusion in poultry diets has resulted in positive effects on growth rate, feed efficiency and egg quality (Moyo and Moyo, 2022). The insect species used for making insect meal are found in abundance in tropical climates and are reared by some small-scale farmers under natural environmental conditions, making them a cheap source of protein for smallholder farmers. In developing countries, where environmental conditions are not conducive for rearing insects, modern facilities with controlled environments are used, which may affect cost (Gasco et al., 2023).

Insects can be offered to animals as live larvae or as processed feed. The most noticeable improvement in growth performance is found when live larvae are provided at the highest frequency of about four times per day (Ipema et al., 2020). Insects are usually processed using methods such as freezing, heat treatment or exposure to carbon dioxide. Preservation techniques, like drying or freeze-drying, are then employed to prevent spoilage. Dried insects are ground into a fine powder using mechanical processes like grinding or milling or pelleted to improve feed handling, reduce dust and enhance feed conversion efficiency.

3.2 Earthworms

Earthworms play an important role in the ecosystem where they improve nutrient availability to plants, soil structure, drainage and productivity. Harvesting of earthworms should always take account of their importance to soil quality control. Earthworms are also proving to be important in the feed industry where they provide valuable nutrients for animals. Earthworm meal contains a substantial amount of protein, ranging from 58% to 71%, depending on factors such as species, diet and processing methods (Bahadori et al., 2021; Khan et al., 2016). The protein in earthworm meals is considered of good quality due to its balanced amino acid profile, including indispensable amino acids such as lysine, methionine and tryptophan, which are crucial for animal growth, tissue development and protein synthesis. Earthworm meals are also a rich source of minerals, vitamins and beneficial fatty acids like omega-3 and omega-6 fatty acids (Kumlu et al., 2021). The consumption by earthworms of organic matter rich in fibre results in increased dietary fibre content in earthworm meal. This supports gut health and digestive processes in animals. Increased growth and higher-quality meat can be observed in broilers-fed diets containing up to 30% earthworm meal (Bahadori et al., 2021). However, a major drawback of the use of earthworm meal is low dry matter content of between 15% and 20% (Cayot et al., 2009; Pérez-Corría et al., 2019).

3.3 Rumen digesta

Rumen digesta is a byproduct of the content of the rumen, containing microbial protein and undigested feed particles. It is typically a post-slaughter waste product. The contents of the first compartment of a ruminant animal's stomach are rich in nutrients due to the fermentation processes that occur (Ari et al., 2006). While it is not typically used as a direct feed ingredient in non-ruminant diets, understanding its nutritional profile can provide insights into its potential as an alternative protein source. Rumen digesta contains microbial protein produced during the fermentation of ingested feed (Cherdthong, 2020). The protein content and the amino acid profile of rumen digesta are influenced by microbial activity, the composition of the original feed and the stage of digestion. The microbes involved in the fermentation process include bacteria, protozoa and fungi. These microbes can contribute to the nutrient profile, providing protein and other beneficial compounds. Other benefits are the production of volatile fatty acids (VFAs) as the end products of microbial fermentation in the rumen (Elfaki and Abdelatti, 2016). They include acetate, propionate and butyrate, which are a source of energy for the animal. Rumen biodigesta also contains vitamins (B vitamins) and minerals such as potassium, phosphorus and magnesium, which are released from feed materials during fermentation and are important for animal health. Rumen digesta can be collected after slaughter, dried and ground into a meal.

3.4 Tankage

Tankage is the residue from tanks used in rendering of animal carcasses and is rich in protein, minerals and amino acids (Cromwell, 2006). Animal tissues are cooked, dried and ground to create tankage, providing a concentrated source of protein and essential nutrients, typically containing around 50–70% protein. The high protein content makes it a valuable ingredient for promoting growth and development in animals, including poultry (Khan, 2018). Tankage provides a diverse amino acid profile, including essential amino acids required by poultry. It is rich in lipids and minerals such as calcium, phosphorus and trace minerals. Calcium supports bone health while phosphorus is essential for energy metabolism and overall growth. Tankage may contain collagen and gelatin derived from connective tissues and have potential health benefits for animals. The strong odour and flavour of tankage can act as a palatability enhancer in animal diets, encouraging consumption and utilization by poultry. Animal tissues, such as bones and offal, are rendered through cooking and drying. The rendered material is finely ground to create tankage. Proper rendering processes ensure the removal of unwanted materials to meet feed industry standards.

3.5 Blood by-products

Most of the blood used to make blood by-products is sourced from abattoirs during the slaughter of livestock. The composition of blood is comparable across species, consisting of 40% blood cells and 60% plasma (Kuan et al., 2018). Blood meal is mainly processed from fresh whole blood, whereas blood plasma meal is from blood plasma. For blood meal, fresh clean blood can be dried using different methods like spray-drying, flash drying and drum drying (Hardy and Kaushik, 2021). Blood plasma, on the other hand, is created by treating blood with an anticoagulant to stop the clotting process (Makara et al., 2016). Plasma and blood cells are separated from one another. The resulting solution of plasma is converted into a powder using spray drying. Both blood and blood plasma meals are rich in protein and essential amino acids (Table 1). However, nutrients in blood plasma meal are highly digestible compared to nutrients in blood meal. The dietary inclusion of spray-dried porcine plasma in broiler chickens challenged with *Salmonella sofia* improved performance, reduced the negative physiological effects of the pathogen and enhanced immune response (Beski et al., 2015). Spray-dried porcine plasma can be seen to improve overall health and well-being in poultry.

4 Potential alternative sources of protein: plant sources

The use of alternative protein sources for all classes of poultry has become more necessary because, like other non-ruminant animals, poultry requires

Table 1 Nutrient composition of blood by-products

	Blood meal[1]	Plasma meal[2]
Nutrients (g/kg)		
Dry matter	935.7	920
Crude protein	904	780
Crude fat	3.7	3
Ash	245	100
Metabolisable energy, poultry (MJ/kg)	–	15.99
Amino acid (total) (g/kg)		
Arginine	39.4	47
Lysine	86.0	68
Methionine	15.5	7
Cysteine	11.8	28
Methionine + cysteine	–	35
Tryptophan	40.7	14
Glycine	40.4	30
Histidine	54.0	28
Leucine	109.5	78
Isoleucine	11.6	29
Phenylalanine	64.1	4.6
Threonine	40.7	48
Valine	83.3	9
Minerals (g/kg)		
Calcium	2.6	1.5
Sodium	–	22
Phosphorus (available)	–	13
Phosphorus (total)	16	13
Chloride	–	11
Magnesium	–	0.3
Iron (mg/kg)	–	90

[1]Liu et al., 1989.
[2]Beski et al. 2015.

high-quality protein sources in their diets for optimal growth and development. Several oilseeds and leaf meals have been identified as potential replacements for conventional oilseeds as protein meals in poultry feed given their nutritional, environmental and cost benefits. Leaf meals, seaweed, algae and yeast also offer emerging alternative feed resources for poultry. Their utilization enhances nutrient diversity, health and sustainability in poultry feeding. As the industry evolves, these resources contribute to efficient and innovative feed ingredients for poultry nutrition.

4.1 Leaf meals

Leaf meals are derived from various plants and can provide protein, fibre and micronutrients. Leaves are dried and ground to produce leaf meals for poultry (Bamdad et al., 2019; Bakare et al., 2021). The following are some examples.

Moringa: Native to north India, moringa can be grown in tropical and sub-tropical regions of Asia and Africa. Moringa leaves are rich in protein, essential amino acids, vitamins (A, C, B), minerals (calcium, iron) and antioxidants (flavonoids, polyphenols) (Yadav et al., 2024). Leaves provide high-quality protein, contributing to growth and development in poultry (Musa-Azara et al., 2013). Moringa has a high vitamin A content and supports vision, immunity and overall health in poultry. Calcium and iron enhance bone strength and blood health, respectively, in birds, and antioxidants support immune function and protect cells from oxidative stress.

Baobab: Baobab (*Adansonia digitata*) is a widely distributed tree found in Africa. Young leaves of baobab trees can be a potential source of protein (Chadare et al., 2008). They are also rich in essential amino acids, vitamins (A, C, B), minerals (calcium, potassium) and dietary fibre. The high vitamin C content enhances immune function and reduces stress in birds. The available calcium and potassium in baobab leaf meal support bone health, muscle function and electrolyte balance. Dietary fibre in leaf meals promotes gut health, aiding in digestion and nutrient absorption.

4.2 Oilseeds

Oilseed meals are mainly used as primary sources of protein. As noted earlier, there is a need to come up with other seed meals to replace soyabean meal. Some of these oil seed meals with the potential to replace soyabean meal are the Locust Bean seed meal, Baobab seed meal, African Mesquite seed meal and Roselle seed meal.

4.3 Locust bean seed meal

Parkia filicoidea is a leguminous tree found in certain regions of Africa. The meal is obtained by processing the Locust bean seeds, resulting in a protein-rich material. It contains essential amino acids, fibre, vitamins (especially B vitamins), minerals (calcium, phosphorus) and other nutrients.

The protein content of *Parkia filicoidea* seed meal ranges from 25% to 35% (Hassan and Umar, 2005). The amino acid profile of this oilseed includes essential amino acids, with lysine and methionine potentially being limiting (Ari and Ayanwale, 2012). The meal also contains dietary fibre, which can contribute to gut health and digestion. The opportunities for *P. filicoidea* utilization include its wide availability as trees are well-suited to certain African climates, contributing to local sustainability. The high protein content of the seed meal makes it suitable as an alternative protein source, and utilizing *P. filicoidea* seed meal can create economic opportunities in regions where the tree is abundant thus, its agroforestry potential for local economies. The meal contains compounds like tannins and trypsin inhibitors, which can affect nutrient absorption. Processing methods are employed to mitigate these factors.

4.4 Baobab seed meal

Baobab is known for its nutritious fruits and seeds. It is a drought-resistant plant requiring minimal inputs which is not consumed by humans so can be a sustainable source. Baobab seed meal has a protein content ranging from 20% to 30%, making it a suitable potential protein source for monogastric animals (Fatima et al., 2022). It contains a well-balanced amino acid profile, although lysine and methionine may be limiting, requiring supplementation. The baobab seed meal is obtained by removing the seed kernels from the fruit and processing them into a protein-rich meal. The meal is also rich in fibre, vitamins (B-complex and vitamin C), minerals (calcium, potassium and magnesium) and antioxidants. Baobab seed meal can be included in poultry diets as a partial replacement for traditional protein sources with resultant improvement in egg production, weight gain and meat quality (Nxele, 2016). Baobab seeds may contain compounds like tannins and phytates that could limit nutrient absorption. Therefore, processing techniques are required to improve oilseed meal.

4.5 African mesquite seed meal

Prosopis africana, a tree native to Africa, offers potential as an alternative protein source in non-ruminant animal nutrition. Seed meal contains essential amino acids, fibre, vitamins (vitamin B_6), minerals (calcium, potassium) and antioxidants. The protein content ranges from 20% to 28%, and the amino acid profile includes both essential and non-essential amino acids (Akande and Alabi, 2021). *Prosopis africana* trees are drought-resistant and can thrive in arid regions. The utilization of *Prosopis africana* seed meal can contribute to income generation in regions where the tree grows. Amino acid imbalances, especially lysine and methionine are the major setbacks. Similarly, the seed meal may contain compounds like tannins that affect nutrient absorption. Processing methods can mitigate these factors (Yusuf et al., 2013).

4.6 Roselle seed meal

The utilization of Roselle seed meal (*Hibiscus sabdariffa*) as an alternative protein source in the nutrition of non-ruminant animals, especially poultry, is gaining acceptability. The protein content of Roselle seed meal can be close to 40% (Fagbenro et al., 2004). The amino acid profile is well-balanced, although certain amino acids may require supplementation. The meal contains dietary fibre, promoting gut health and digestion. The meal contains essential amino acids, fibre, vitamins (B vitamins and vitamin C), minerals (iron and calcium) and antioxidants. The antinutritional factors associated with Roselle seed meal might contain compounds like tannins that affect nutrient absorption. There are processing methods that can mitigate these factors. Roselle seed meal can be included in poultry diets as a partial protein source with positive effects on growth performance and egg production.

4.7 Algae

Algae are microorganisms rich in protein, omega-3 fatty acids and pigments. The main algae currently used in animal nutrition are *Chlorella vulgaris* (green, single-celled microalgae), *Spirulina platensis* (filamentous and multicellular blue-green algae), *Ulva lactuca* (flat green algae) and *Sargassum* sp (large brown seaweed like algae). The chemical composition and nutritional quality vary with species of algae (Table 2). Algae are harvested, dried, and processed into meal form. Algae offer essential nutrients and bioactive compounds, promoting poultry health and performance. According to Zheng et al. (2012), egg productivity increased from 55.4% in the control group to 59% in 80-week-old Hy-Line layers supplemented with diets containing 2% *Chlorella vulgaris*. The egg yolk colour of 35–43-week-old White Leghorn hens improved after supplementing of diets with 4% or 8% *Sargassum* spp (Carrillo et al., 2012). The addition of algae to broiler diets can also modify the quality of meat.

Table 2 Nutritional composition of algae species

	Chlorella vulgaris	Spirulina platensis	Ulva lactuca	Sargassum sp.
Nutrient composition				
Protein (%)	57	60	19	14
Lipids (%)	20	6	0.3	6
Carbohydrates (%)	23	34	80.7	80
Minerals (mg/kg)				
Na	134.6	189.7	351.7	389.3
K	4.99	132.6	209	244.3
Ca	59.3	88.3	180.7	176
Fe	25.9	10.3	34.5	32.2
Zn	1.2	2.5	1.78	5.8
Cu	0.6	0.32	1.83	1.62
Se	0.7	0.37	1.6	49.8
Mn	0.21	0.38	4.8	3.3

Source: Adapted from: Coudert et al. (2020).

5 Improving the nutritional value of alternative protein sources for poultry: chemical, biological and physical treatments

Feed for poultry can be made from a variety of feed ingredients. However, the nutritional value of these feed ingredients varies. Ingredients have different physicochemical properties, which make some of the nutrients unavailable for use by the birds. Some of the properties of ingredients that prevent utilization of the nutrients are plant fibres, plant secondary metabolites and waxes in plants and insects (Meremikwu et al., 2013; Nascimento et al., 2020, 2021). Plant secondary compounds and waxes prevent the use of several nutrients and minerals (Soetan et al., 2009). In addition, some plant secondary compounds cause toxicity to animals (Tadele, 2015). Over the past 20 years, much work has been done to improve the nutritional value of alternative feed ingredients. The nutritional value of the ingredients can be improved by using methods that fall into three categories: chemical, biological and physical treatment.

5.1 Chemical treatment

Chemical treatments use substances that can alter molecular structures through chemical interactions. The treatment processes involve the use of alkalis such as sodium hydroxide, ammonium hydroxide, and sodium bicarbonate or reducing agents include the use of sodium metabisulfite, metabisulfite, *tris* (2-carboxyethyl-phosphine [TCEP]) (Guerrero-Beltran et al., 2009; Avilés-Gaxiola et al., 2018). Most proteins in plants are prevented from being utilised by animals because of trypsin inhibitors (TIs). TIs inhibit enzyme activity by blocking the active site of the enzymes (Cristina et al., 2019). The N- or C-terminus and the exposed loop of protease inhibitors are frequently considered critical structural features for inhibiting enzyme activity. Acids, bases and reducing agents help by cleaving the disulphide bond and making proteins available for use by poultry. A study by Avilés-Gaxiola et al. (2018) found that the temperature required to deactivate TI in soybeans is lowered from 100 to 90.5°C and 73.9°C, respectively, when cooked in the presence of 1% NaOH or NH_4OH. Likewise, soyabeans in 0.1M $Na_2S_2O_5$ aqueous solution boiled for 2 h at 65°C inactivated trypsin inhibitor by more than 94%. Most leguminous tree species produce seeds that can be used as feed for animals (Mehari et al., 2019). Examples of some of these leguminous trees are *Laecana leucociphala*, *Calliandra*, and *Acacia* spp. The seeds are rich in crude protein and can also be processed the same way as soya beans to make proteins available for use by poultry.

Insect meal is now a popular protein ingredient in some countries. The exoskeleton of insects contains a cuticle consisting of alternate layers of protein and chitin impregnated with calcium carbonate and pigments and interspersed with polyphenols (Muthukrishnan et al., 2020). Sclerotin is the form of protein found in the exoskeleton of insects. It is formed by cross-linking classes of protein molecules. Acids, bases and reducing agents also help break down the complexes and make protein available for use by poultry (Pellis et al., 2022).

Apart from acids and bases, polyethylene glycol (PEG) can also be used to make proteins available for use by animals. Polyethylene glycol is a synthetic polymer poorly digested and absorbed in the intestines of animals (Grosell and Genz, 2006). It has a high affinity with tannins and can bind with tannins to form a tannin–PEG complex (Makkar et al., 1995). A study by Ngwa et al. (2002) reported the use of PEG to improve crude protein digestibility in sheep. Similarly, Xie et al. (2021) found that adding PEG to diets containing high tannins improved CP digestibility. Van Niekerk et al. (2020) found the use of PEG to inactivate tannins and hence improved the utilization of protein and the overall performance of broilers. Polyethylene glycol may, therefore, be used for novel ingredients high in tanniferous compounds when formulating diets for poultry.

5.2 Biological treatment

Different biological methods are used to inactivate trypsin inhibitors and make protein available for animal use. Examples of some of the methods are germination and fermentation (Nkhata et al., 2018). All plants naturally go through the biological process of germination. In this process, the seed emerges from dormancy, where some quantitative and qualitative changes occur. These changes depend on plant species, plant variety and germination conditions. During germination, seed enzymatic systems are activated, protease activity increases, protein nitrogen decreases and peptides, polypeptides and non-protein amino acids increase (Bamdad et al., 2009). The production of proteases is believed to inactivate trypsin inhibitors and make proteins available for use by the animal. A study by Sangronis and Machado (2007) found that germination increased *in vitro* protein digestibility by 2% (white beans), 3% (black beans) and 4%.

Fermentation is another biological method used to preserve feed for livestock. There are so many physicochemical changes that occur during the fermentation process. Proteolysis is a process where proteins are broken down into individual amino acids (Fijałkowska et al., 2015). Animals can then easily assimilate the amino acids. Anti-nutritional factors have also been reported to decrease because of anaerobic fermentation (Kumar et al., 2022). The fermentation process can be improved by additives such as probiotics, urea and molasses to mention a few (Kholif et al., 2022). Enzymes released by the probiotics facilitate the breaking down of fibres and proteins. A plethora of studies have used this method to ensile forage for ruminants (Keady et al., 2013; Evans, 2018; Campbell et al., 2020). Lately, this method has also been used to improve forage to be used by non-ruminant animals. Cassava leaves, a potential alternative protein source, can be processed using this method (Bhavna et al., 2023).

5.3 Physical/mechanical treatment

Physical and/or mechanical treatment methods are used to process feed for animal feeding. The methods have an advantage in that they destroy pathogenic microorganisms, increase the palatability of feed, make the storage easy and safe, increase nutrient content and availability, and change the particle size or density of the feed. Dry heat and hydrothermal treatment are classified under physical/mechanical treatment. The methods reduce anti-nutritional factors in plants, including tannins and trypsin inhibitors, which prevent protein utilization (Ojo, 2022). According to Kadam and Smithard (1987), 80% and 90% of the inhibitors in soybeans are inactivated when boiled for 14 min and for 30 min, respectively. The same methods can be used for treating non-conventional protein sources.

Heat is used to dehydrate the insects, before grinding them into powder and incorporating into feed. Heat is well known for altering the structure of proteins, thereby exposing sites for enzyme action to facilitate protein digestion by animals. Many of these treatments are not practical to use in commercial farming settings if electric power from grid lines is used, as they require large machines for processing. As a result, these treatments are economically unprofitable for large-scale commercial farms as the benefits may be too low or even negative.

6 Environmental sustainability of using alternative protein sources feed for poultry

Feed production has negative and positive effects on the environment, both direct and indirect. For instance, cultivating feed crops adds to the eutrophication and acidification of the environment, primarily due to nitrate leaching into the water and NH_3 emissions into the atmosphere. Chemicals like herbicides, pesticides and fertilisers used during the cultivation of crops all contribute to the pollution of the environment (Baweja et al., 2020). Insect meals, on the other hand, can be reared easily at an economically and environmentally sustainable cost. For example, black soldier fly larvae and maggots can be reared on substrates such as manure and food waste (Makkar et al., 2014). Black soldier fly larvae and maggots can convert organic waste into valuable biomass (van Huis et al., 2013), solving several environmental problems associated with manure and other organic wastes.

Poultry manure management practices can contribute to reducing environmental pollution. The amount of nitrogen excreted depends on the birds' ability to convert feed into energy and the diet's protein content. Reducing the feed-related impacts of poultry production on the environment can be done in two ways. First, improving the feed efficiency, i.e. reducing the amount of feed needed for a particular body weight gain or egg production, would reduce the emissions from feed production and manure management (Oxenboll et al., 2011). Second, selecting feed ingredients with lower environmental impacts during the production stage or more balanced nutrient content should be possible, reducing the excretion of nutrients such as nitrogen and phosphorus (Leinonen and Kyriazakis, 2016). In an ideal situation, diets that fulfil both criteria would be expected to produce a maximal reduction of the environmental impacts. Studies have shown insect meals to be highly digestible. Insect meals (housefly larvae) have up to 98.5% CP digestibility; hence, lower nitrogen losses, and lower eutrophication levels (Hwangbo et al., 2009).

7 Consumer perceptions of feeding animals with alternative feed sources

There has been a growing interest in the consumption of organic food (Frewer et al., 2003). When it comes to meat consumption, consumers would like to know how the animals were reared and how they were fed. Most people are now leaning towards organic feed, which is not linked to some of the chronic ailments in the human population. Feeding chickens leaf meals will likely make poultry meat more acceptable by consumers. The meat is likely to be rejected if there are changes to sensory characteristics of the meat. A plethora of studies have investigated the use of different novel ingredients on the sensory characteristics of meat. Ncube et al. (2018) fed acacia leaves to broiler chickens with no adverse effects on of the organoleptic properties of meat. Similarly, meat quality attributes were not affected by feeding broiler chickens cassava leaves (Bakare et al., 2021). Feeding grasshopper meal had no discernible effects on the sensory panel results for breast and thigh meats (Khan, 2018). There is, therefore, a potential for using alternative plant and insect protein sources.

Human entomophagy, also known as eating insects, is a widespread and culturally ingrained practice of many populations in developing countries in Central and South America, Africa and Asia. Some examples of insects that form a delicacy in these countries include crickets, cicadas, grasshoppers, ants, mealworms (larvae of the darkling beetle) and certain types of caterpillars like mopane worms (Bernard and Womeni, 2017). In countries where insects have long served as traditional foods, they can also be used as feed for livestock when in abundance. For example, in some countries experiencing locust outbreaks, farmers take advantage of this by harvesting the locusts and feeding them to animals (Cerritos and Cano-Santana, 2008). Consumption of products from animals fed insect meals is highly acceptable in these countries. For example, in the Philippines, free-range chickens fed on grasshoppers were found to taste better compared to those fed on feed with conventional protein sources (Khusro et al., 2012). This is not the case in Western societies where entomophagy is not embedded in culinary traditions (Ghosh et al., 2018). Consumption of insects in some of the Western countries is regarded as unacceptable by most of the human populace (Wendin and Nyberg, 2021).

8 Conclusion

Emerging protein sources will play a strong role in poultry feeding in the coming decades. This is mainly due to the result of recurring constraints in the supply of conventional sources due to climate change or political upheavals in major areas of production. It is likely that many of the emerging sources will be produced in commercial quantities in line with the general growth of agriculture in many developing countries. Such a change will have only minimal impact on the prices of conventional sources, which will continue to be used by major producers around the world.

9   References

Akande, K. E. and Alabi, O. J. 2021. The utilization of African mesquite (Prosopis africana) as potential feedstuff for monogastric animals: a review. *Nigerian Journal of Animal Science* 23(1), pp.168–172.

Ari, M. M. and Ayanwale, B. A. 2012. Nutrient retention and serum profile of broilers fed fermented African Locust beans (Parkia filicoide). *Asian Journal of Agricultural Research* 6(3), pp.129–136.

Ari, M. M., Maikaffi, Y. M., Barde, R. E., Aya, V. E. and Ogah, D. M. 2006. Effect of biostatic agents on the short-term preservation of wet rumen content. *PAT* 2(1).

Avilés-Gaxiola, S., Chuck-Hernández, C. and Serna Saldívar, S. O. 2018. Inactivation methods of trypsin inhibitor in legumes: a review. *Journal of Food Science* 83(1), pp.17–29.

Bahadori, Z., Esmaylzadeh, L. and Torshizi, M. A. K. 2021. The effect of earthworm meal on growth performance, immunity, intestinal microbiota, carcass characteristics and meat quality of broiler chickens. *Livestock Science* 202:74–81.

Bakare, A. G., Cawaki, P., Ledua, I., Bautista-Jimenez, V., Kour, G., Sharma, A. C. and Tamani, E. 2021. Quality evaluation of breast meat from chickens fed cassava leaf meal-based diets. *Animal Production Science* 61(6), pp.613–619.

Bamdad, F., Dokhani, S. and Keramat, J. 2009. Functional assessment and subunit constitution of Lentil (lens culinaris) proteins during Germination. *International Journal of Agriculture and Biology* 11(6), pp.690–694.

Baweja, P., Kumar, S. and Kumar, G. 2020. Fertilizers and pesticides: -their impact on soil health and environment. In Giri, B. and Varma, A. (eds). Soil health, Springer Nature, Geneva, Switzerland (pp.265–285). Cham: Springer.

Bernard, T. and Womeni, H. M. 2017. Entomophagy: insects as food. *Insect Physiology and Ecology* 2017, pp.233–249.

Beski, S. S. M., Swick, R. A. and Iji, P. A. 2015. The effect of the concentration and feeding duration of spray-dried plasma protein on growth performance, digestive enzyme activities, nutrient digestibility and intestinal mucosal development of broiler chickens. *Animal Production Science* 56(11), pp.1820–1827.

Bhavna, A., Zindove, T. J., Iji, P. A. and Bakare, A. G. 2023. Growth performance, carcass characteristics and meat sensory evaluation of broiler chickens fed diets with fermented cassava leaves. *Animal Bioscience.* https://doi.org/10.5713/ab.23.0362.

Campbell, M., Ortuño, J., Ford, L., Davies, D. R., Koidis, A., Walsh, P. J. and Theodoridou, K. 2020. The effect of ensiling on the nutritional composition and fermentation characteristics of brown seaweeds as a ruminant feed ingredient. *Animals* 10(6), p.1019.

Carrillo, S., Bahena, A., Casas, M., Carranco, M. E. and Calvo, C. C. 2012. The alga Sargassum spp. as alternative to reduce egg cholesterol content. *Cuban Journal of Agricultural Science* 46(2), pp.1–6.

Cayot, N., Cayot, P., Bou-Maroun, E., Laboure, H., Abad-Romero, B., Pernin, K., Seller-Alvarez, N., Hernández, A. V., Marquez, E. and Medina, A. L. 2009. Physico-chemical characterisation of a non-conventional food protein source from earthworms and sensory impact in arepas. *International Journal of Food Science & Technology* 44(11), pp.2303–2313.

Cerritos, R. and Cano-Santana, Z. 2008. Harvesting grasshoppers *Sphenarium purpurascens* in Mexico for human consumption: a comparison with insecticidal control for managing pest outbreaks. *Crop Protection* 27(3–5), pp.473–480.

Chadare, F. J., Linnemann, A. R., Hounhouigan, J. D., Nout, M. J. R. and Van Boekel, M. A. J. S. 2008. Baobab food products: a review on their composition and nutritional value. *Critical Reviews in Food Science and Nutrition* 49(3), pp.254–274.

Cherdthong, A. 2020. Potential use of rumen digesta as ruminant diet–a review. *Tropical Animal Health and Production* 52(1), pp.1–6.

Coudert, E., Baéza, E. and Berri, C. 2020. Use of algae in poultry production: a review. *World's Poultry Science Journal* 76(4), pp.767–786.

Cristina Oliveira de Lima, V., Piuvezam, G., Leal Lima Maciel, B. and Heloneida de Araújo Morais, A. 2019. Trypsin inhibitors: promising candidate satietogenic proteins as complementary treatment for obesity and metabolic disorders? *Journal of Enzyme Inhibition and Medicinal Chemistry* 34(1), pp.405–419.

Cromwell, G. L. 2006. Rendered products in swine nutrition. In D. L. Meeker (ed.), *Essential Rendering*, p.141. Alexandria, VA: Renderers Association.

Elfaki, M. O. and Abdelatti, K. A. 2016. Rumen content as animal feed: a review. *Journal of Veterinary Medicine and Animal Production* 7(2), pp.80–88.

Evans, B. 2018. The role ensiled forage has on methane production in the rumen. *Animal Husbandry, Dairy and Veterinary Science* 2(4), pp.1–4..

Fagbenro, O. A., Akande, T. T., Fapohunda, O. O. and Akegbejo-Samsons, Y. 2004. Comparative assessment of roselle (Hibiscus sabdariffa var. sabdariffa) seed meal and kenaf (Hibiscus sabdariffa var. altissima) seed meal as replacement for soybean meal in practical diets for fingerlings of Nile tilapia Oreochromis niloticus. In *6th International Symposium on Tilapia in Aquaculture*, pp.277–287. Manila, Philippines: Philippine International Convention Center Roxas Boulevard.

Fatima, G. A., Hamdon, A. A. and Mekki, D. M. 2022. Impacts of feeding graded levels of baobab seed meal on growth performance and feed utilization of broiler chicks. *Acta Sci. Pharma.* 3(10): 2–7.

Fijałkowska, M., Pysera, B., Lipiński, K. and Strusińska, D. 2015. Changes of nitrogen compounds during ensiling of high protein herbages–a review. *Annals of Animal Science* 15(2), pp.289–305.

Fraanje, W. and Garnett, T. 2020. *Soy: food, feed and land use change* (Foodsource: Building Blocks). Oxford, UK: Food Climate Research Network, University of Oxford.

Frewer, L., Scholderer, J. and Lambert, N. 2003. Consumer acceptance of functional foods: issues for the future. *British Food Journal* 105(10), pp.714–731.

Gasco, L., Renna, M., Bellezza Oddon, S., Rezaei Far, A., Naser El Deen, S. and Veldkamp, T. 2023. Insect meals in a circular economy and applications in monogastric diets. *Animal Frontiers* 13(4), pp.81–90.

Ghosh, S., Jung, C. and Meyer-Rochow, V. B. 2018. What governs selection and acceptance of edible insect species? In A. Halloran et al. (eds.), *Edible insects in sustainable food systems*, pp.331–351. Cham: Springer Publication.

Grosell, M. and Genz, J. 2006. Ouabain-sensitive bicarbonate secretion and acid absorption by the marine teleost fish intestine play a role in osmoregulation. *American Journal of Physiology-Regulatory, Integrative and Comparative Physiology* 291(4), pp.R1145–R1156.

Guerrero-Beltrán, J. A., Estrada-Girón, Y., Swanson, B. G. and Barbosa-Cánovas, G. V. 2009. Pressure and temperature combination for inactivation of soymilk trypsin inhibitors. *Food Chemistry* 116(3), pp.676–679.

Hardy, R. W. and Kaushik, S. J. eds. 2021. *Fish nutrition*. San Diego, CA: Academic Press.

Hassan, L. G. and Umar, K. J. 2005. Protein and amino acids composition of African locust bean (Parkia biglobosa). *Tropical and Subtropical Agroecosystems* 5(1), pp.45–50.

Hwangbo, J., Hong, E. C., Jang, A., Kang, H. K., Oh, J. S., Kim, B. W. and Park, B. S. 2009. Utilization of house fly-maggots, a feed supplement in the production of broiler chickens. *Journal of Environmental Biology* 30(4), pp.609–614.

Ipema, A. F., Gerrits, W. J., Bokkers, E. A., Kemp, B. and Bolhuis, J. E. 2020. Provisioning of live black soldier fly larvae (Hermetia illucens) benefits broiler activity and leg health in a frequency-and dose-dependent manner. *Applied Animal Behaviour Science* 230, p.105082.

Jedrejek, D. et al. 2016. Animal by-products for feed: characteristics, European regulatory framework, and potential impacts on human and animal health and the environment. *Journal of Animal and Feed Sciences* 25, pp.189–202.

Kadam, S. S. and Smithard, R. R. 1987. Effects of heat treatments on trypsin inhibitor and hemagglutinating activities in winged bean. *Plant Foods for Human Nutrition* 37, pp.151–159.

Keady, T., Hanrahan, S., Marley, C. and Scollan, N. D. 2013. Production and utilization of ensiled forages by beef cattle, dairy cows, pregnant ewes and finishing lambs-A review. *Agricultural and Food Science* 22(1), pp.70–92.

Khan, S., Naz, S., Sultan, A., Alhidary, I. A., Abdelrahman, M. M., Khan, R. U., Khan, N. A., Khan, M. A. and Ahmad, S. 2016. Worm meal: a potential source of alternative protein in poultry feed. *World's Poultry Science Journal* 72(1), pp.93–102.

Khan, S. H. 2018. Recent advances in role of insects as alternative protein source in poultry nutrition. *Journal of Applied Animal Research* 46(1), pp.1144–1157.

Kholif, A. E., Hamdon, H. A., Gouda, G. A., Kassab, A. Y., Morsy, T. A. and Patra, A. K. 2022. Feeding date-palm leaves ensiled with fibrolytic enzymes or multi-species probiotics to Farafra ewes: Intake, digestibility, ruminal fermentation, blood chemistry, milk production and milk fatty acid profile. *Animals* 12(9), p.1107.

Khusro, M., Andrew, N. R. and Nicholas, A. 2012. Insects as poultry feed: a scoping study for poultry production systems in Australia. *World's Poultry Science Journal* 68(3), pp.435–446.

Kouřimská, L. and Adámková, A. 2016. Nutritional and sensory quality of edible insects. *NFS Journal* 4, pp.22–26.

Kuan, D. H., Wu, C. C., Su, W. Y. and Huang, N. T. 2018. A microfluidic device for simultaneous extraction of plasma, red blood cells, and on-chip white blood cell trapping. *Scientific Reports*, 8(1), p.15345.

Kumar, Y., Basu, S., Goswami, D., Devi, M., Shivhare, U. S. and Vishwakarma, R. K. 2022. Anti-nutritional compounds in pulses: implications and alleviation methods. *Legume Science* 4(2), p.e111.

Kumlu, M., Beksari, A., Yilmaz, H. A., Sariipek, M., Kinay, E., Turchini, G. and Eroldogan, O. T. 2021. n-3 LC-PUFA enrichment protocol for red earthworm, Eisenia fetida: a cheap and sustainable method. *Turkish Journal of Fisheries and Aquatic Sciences*. https://doi.org/10.4194/1303-2712-v21_7_03.

Leinonen, I. and Kyriazakis, I. 2016. How can we improve the environmental sustainability of poultry production? *Proceedings of the Nutrition Society* 75(3), pp.265–273.

Liu, J. K., Waibel, P. E. and Noll, S. L. 1989. Nutritional evaluation of blood meal and feather meal for turkeys. *Poultry Science* 68(11), pp.1513–1518.

Makara, A., Kowalski, Z., Fela, K. and Generowicz, A. 2016. Utilization of animal blood plasma as example of using cleaner technologies methodology. *Technical Transactions/Czasopismo Techniczne* 11. https://doi.org/10.4467/2353737XCT.16.197.5946.

Makkar, H. P., Blümmel, M. and Becker, K. 1995. In vitro effects of and interactions between tannins and saponins and fate of tannins in the rumen. *Journal of the Science of Food and Agriculture* 69(4), pp.481–493.

Makkar, H. P., Tran, G., Heuzé, V. and Ankers, P. 2014. State-of-the-art on use of insects as animal feed. *Animal Feed Science and Technology* 197, pp.1–33.

Mehari, R., Shumuye, B., Kinfe, M. and Adugna, G. 2019. Performance of egg-laying hens fed Acacia saligna seed meal. Livestock Research for Rural Development 31: Article 27.

Meremikwu, V. N., Ibekwe, H. A. and Essien, A. 2013. Improving broiler performance in the tropics using quantitative nutrition. *World's Poultry Science Journal* 69(3), pp.633–638.

Moyo, S. and Moyo, B. 2022. Potential utilization of insect meal as livestock feed. In *Animal feed science and nutrition-production, health and environment*. IntechOpen.

Musa-Azara, S. I., Jibrin, M., Ari, M. M., Hassan, D. I. and Ogah, D. M. 2014. Effects of Moringa oleifera Linn seed administration on sperm production rate and gonadal sperm reserve in rabbits. *British Biotechnology Journal* 4(7), p.801.

Musa-Azara, S. I., Ogah, D. M., Yakubu, A., Ari, M. M. and Hassan, D. I. 2013. Effects of Hibiscus calyx extracts on the blood chemistry of broiler chickens. *Egyptian Poultry Science* 33(1), pp.309–312.

Muthukrishnan, S., Mun, S., Noh, M. Y. Geisbrecht, E. R. and Arakane, Y. 2020. Insect cuticular chitin contributes to form and function. *Current Pharmaceutical Design* 26(29), pp.3530–3545.

Nascimento, M. Q. D., Gous, R. M., Reis, M. P., Viana, G. S., Nogueira, B. R. F. and Sakomura, N. K. 2021. Gut capacity of broiler breeder hens. *British Poultry Science* 62(5), pp.710–716.

Nascimento, M. Q. D., Gous, R. M., Reis, M. D. P., Fernandes, J. B. K. and Sakomura, N. 2020. Prediction of maximum scaled feed intake in broiler chickens based on physical properties of bulky feeds. *British Poultry Science* 61(6), pp.676–683.

Ncube, S., Halimani, T. E., Chikosi, E. V. I. and Saidi, P. T. 2018. Effect of Acacia angustissima leaf meal on performance, yield of carcass components and meat quality of broilers. *South African Journal of Animal Science* 48(2), pp.271–283.

Ngwa, A. T., Nsahlai, I. V. and Iji, P. A. 2002. Effect of supplementing veld hay with a dry meal or silage from pods of Acacia sieberiana with or without wheat bran on voluntary intake, digestibility, excretion of purine derivatives, nitrogen utilization, and weight gain in South African Merino sheep. *Livestock Production Science* 77(2–3), pp.253–264.

Nkhata, S. G., Ayua, E., Kamau, E. H. and Shingiro, J. B. 2018. Fermentation and Germination improve nutritional value of cereals and legumes through activation of endogenous enzymes. *Food Science & Nutrition* 6(8), pp.2446–2458.

Nxele, S. K. 2016. *The Potential Use of Baobab (Adansonia Digitata) Seedcake as a Commercial Diet Replacement and in Feed Formulation for Broiler Chickens* (Doctoral dissertation). University of Fort Hare.

Ojo, M. A. 2022. Tannins in foods: nutritional implications and processing effects of hydrothermal techniques on legume seeds. *Prev. Nutr. Food Sci.* 27(1): 14–19.

Øverland, M., Mydland, L. M. and Skrede, A. 2017. Marine macroalgae as a source of protein and bioactive compounds in feed for non-ruminant animals. *Journal of the Science of Food and Agriculture* 99(1). https://doi.org/10.1002/jsfa.9143.

Oxenboll, K. M., Pontoppidan, K. and Fru-Nji, F. 2011. Use of a protease in poultry feed offers promising environmental benefits. *International Journal of Poultry Science* 10(11), pp.842–848.

Pellis, A., Guebitz, G. M. and Nyanhongo, G. S. 2022. Chitosan: sources, processing and modification techniques. *Gels*,8(7), p.393.

Pereira, A. G., Fraga-Corral, M., Garcia-Oliveira, P., Otero, P., Soria-Lopez, A., Cassani, L., Cao, H., Xiao, J., Prieto, M. A. and Simal-Gandara, J. 2022. Single-cell proteins obtained by circular economy intended as a feed ingredient in aquaculture. *Foods* 11(18), p.2831.

Pérez-Corría, K., Botello-León, A., Mauro-Félix, A. K., Rivera-Pineda, F., Viana, M. T., Cuello-Pérez, M., Botello-Rodríguez, A. and Martínez-Aguilar, Y. 2019. Chemical composition of earthworm (Eisenia foetida) co-dried with vegetable meals as an animal feed. *Ciencia y Agricultura* 16(2), pp.79–92.

Sajid, Q. U. A., Asghar, M. U., Tariq, H., Wilk, M. and Płatek, A. 2023. Insect meal as an alternative to protein concentrates in poultry nutrition with future perspectives (An updated review). *Agriculture* 13(6), p.1239.

Sangronis, E. and Machado, C. J. 2007. Influence of Germination on the nutritional quality of Phaseolus vulgaris and Cajanus cajan. *LWT-Food Science and Technology* 40(1), pp.116–120.

Soetan, K. O. and Oyewole, O. E. 2009. The need for adequate processing to reduce the anti-nutritional factors in plants used as human foods and animal feeds: a review. *African Journal of food Science* 3(9), pp.223–232.

Spranghers, T., Ottoboni, M., Klootwijk, C., Ovyn, A., Deboosere, S., De Meulenaer, B., Michiels, J., Eeckhout, M., De Clercq, P. and De Smet, S. 2017. Nutritional composition of black soldier fly (Hermetia illucens) prepupae reared on different organic waste substrates. *Journal of the Science of Food and Agriculture* 97(8), pp.2594–2600.

Tadele, Y. 2015. Important anti-nutritional substances and inherent toxicants of feeds. *Food Science and Quality Management* 36, pp.40–47.

Van Huis, A., Van Itterbeeck, J., Klunder, H., Mertens, E., Halloran, A., Muir, G. and Vantomme, P. 2013. *Edible insects: future prospects for food and feed security* (No. 171). Rome: Food and Agriculture Organization of the United Nations.

Van Niekerk, R. F., Mnisi, C. M. and Mlambo, V. 2020. Polyethylene glycol inactivates red grape pomace condensed tannins for broiler chickens. *British Poultry Science* 61(5), pp.566–573.

Wendin, K. M. and Nyberg, M. E. 2021. Factors influencing consumer perception and acceptability of insect-based foods. *Current Opinion in Food Science* 40, pp.67–71.

Xie, B. et al. 2021. Adding polyethylene glycol to steer rations containing sorghum tannins increases crude protein digestibility. *Animal Nutrition* 7(3): 779–786.

Yadav, L. P., Gangadhara, K., Apparao, V. V., Singh, A. K., Rane, J., Kaushik, P., Sekhawat, N., Malhotra, S. K., Rai, A. K., Yadav, S. L. and Berwal, M. K. 2024. Nutritional, antioxidants and protein profiling of leaves of Moringa oleifera germplasm. *South African Journal of Botany* 165, pp.443–454.

Yusuf, N. D., Ogah, D. M., Hassan, D. I., Musa, M. M., Ari, M. M. and Doma, U. D. 2013. Carcass Evaluation of Broilers fed on Decorticated Fermented Prosopis africana G. seed meal. *Macedonian Journal of Animal Science* 3(1), pp.45–48.

Zheng, L. et al. 2012. The dietary effects of fermented *Chlorella vulgaris* on production performance, liver lipids and intestinal microflora in laying hens. *Asian-Austral. J. Anim. Sci.* 25(2): 261–266.

Chapter 5

High protein corn fermentation products for poultry derived from corn ethanol production

Peter E.V. Williams, FluidQuipTechnologies, USA

1 Introduction

Crops are a primary source of protein, and for many years, soybean meal (SBM) has been the protein of choice for livestock feed. However, the pleiotropic relationship between nitrogen and protein content has, over time, resulted in the reduction in protein content of crops such as soybean and corn (maize) as the economics of grain production have prioritized increased yield. This, in turn, has exacerbated the need for higher-concentration protein supplements for feed formulation. In addition, there has also been a movement to reduce reliance on soy products and replace them with more sustainable sources of protein, particularly vegetable proteins with the absence of anti-nutritional factors.

Decisions on how a feed is formulated are already being driven by the sustainability characteristics of individual feed ingredients. Sustainability (meeting energy and carbon emissions reduction targets) alongside demographics (e.g. the age of those in the labor market) have been identified by the International Feed Industry Federation as a key global issue facing the industry. Indexes of sustainability, such as green house gas (GHG) and land use change (LUC) metrics, are likely to become two new parameters that will be

http://dx.doi.org/10.19103/AS.2024.0143.24

components of feed specification sheets, in addition to the nutritional data that has been the norm.

To improve sustainability, processed animal proteins are being reintroduced into feed formulations, and work is ongoing to produce a range of single-cell proteins, insect protein, and microbial protein. Attention is also being paid to alternative crops high in protein. However, there are considerable challenges to growing new alternative crops on any scale. It takes several years to breed an elite variety to achieve maximum production in different geographic locations with different environmental conditions. Logistical factors such as acquiring appropriate machinery and inputs, as well as acquiring agronomic knowledge to successfully grow a new crop, all serve to favor the status quo. In addition, new crops require land and resources and potentially compete with existing crop production, contributing to GHG emissions from agriculture as well as indirectly exacerbating climate change from changes in land use (through deforestation). In contrast, plant-based co-products, secondary products generated during a manufacturing process, have a unique position in that they do not demand additional acreage and do not compete with human food consumption (Mottet et al., 2017).

2 Distillers dried grains and solubles

Corn (or maize) is the major cereal crop grown in the USA and widely cultivated globally. Traditionally considered an energy crop primarily grown for the energy produced in the form of starch and oil, corn has a limited amount of protein. In addition to its use as a food crop, corn has also become an important feedstock for the sustainable biofuel industry, with over a 100 million tons of corn annually processed into ethanol biofuel in the USA. A by-product of ethanol production is distillers dried grains and solubles (DDGS). This is the dried residue remaining after the starch fraction of corn is fermented with yeasts to produce ethanol. Following fermentation, the ethanol is removed by distillation, and the remaining fermentation residues are dried to become DDGS. The growth in ethanol production from corn has resulted in a plentiful supply of DDGS as a co-product, which has found a new market as a livestock feed. However, it is important to note that DDGS, as a by-product of the dry grind ethanol industry, was never designed as a livestock feed but emerged as a convenient means of marketing the residue from the ethanol production process. This creates a potential opportunity to further develop DDGS as a livestock feed product.

More broadly, the dry grind ethanol industry benefits from several important features as a source of animal feed. A dry grind ethanol plant has established logistics for delivery and dispatch of grain feedstock and any feed products. Individual dry grind ethanol plants benefit from economies of scale, with large plants processing in excess of one million tons of grain per annum. Ethanol plants are also integrated bioprocessing facilities, typically with integrated heat and power generation. These facilities are prime sites for the installation of processing technology to produce new products with higher nutritional value than traditional DDGS products.

3 Corn-fermented protein

There are currently four different commercial processes for the production of high-concentration corn-fermented protein (CFP) products (approximately >50% crude protein) (Fig. 1). By producing products with higher levels of protein compared to conventional DDGS, these technologies offer the potential to diversify and expand the revenue streams of the bioethanol plants, in addition to providing new protein products for the feed industry.

The four processes are an evolution of the wet grind milling process and involve the separation of protein and fiber to produce separate high-value concentrated streams of protein and fiber. Mechanically separated CFP is produced from whole stillage when portions of fiber and oil are removed, concentrating residual grain proteins and yeast by methods involving centrifugation and washing. One version of CFP processing involves the separation of all fractions post fermentation, with all components of the product exposed to fermentation: the Maximized Stillage Co-products (MSC™) process. A second version of CFP manufacture incorporates pre-fermentation grinding and separation of the whole grain. A third recovers the protein via flocculation, and a fourth employs electrostatic forces to separate fiber from protein. Key differences between these processes are identified in Fig. 1.

		Pre Fermentation separation	Fermented	Yeast	Approx crude protein %
Empyreal		Protein concentrate			75
CFP mechanically separated	SEQUENCE™				60
	GPRE NexPro BP50 A+Pro AltiPro				>48
CFP flocculant separated	ProCap Gold				52
CFP	ANDVantage				40
	ANDVantage				50
	PROTOMAX				50

	full exposure to >50 fermentation
	partial exposure to fermentation
	approx 25% spent yeast in dry matter
	undefined yeast content

Figure 1 Technologies for producing a range of commercial high-protein feed products.

The MSC™ process (Fig. 2) separates the protein from the whole stillage by mechanical centrifugation and alternate washing steps. An important element to grasp is that the dry grind ethanol process starts with a 50–70 h fermentation of ground corn. The ground grain is exposed to moisture, raised temperature, and enzymatic hydrolysis in order to facilitate the conversion of starch into ethanol. There has been little focus on the impact of the fermentation process on residual components of the grain, whether fibrous components found in whole stillage or the composition of protein components. Other than testing of phytase for example as a means of potentially increasing protein recovery, the potential value of fermentation in improving product functionality is unexplored. Xu et al. (2020) listed such potential improvements as improved protein digestibility, increased number of small peptides, increased energy digestibility, reduced fiber levels, improved nutritional value, bioavailability, weight gain, and feed conversion rate. Work is in progress to investigate whether additional functional benefits can be obtained for CFP products via the use of fermentation.

In addition, little attention has been given to the fact that, prior to fermentation, there is a significant *in situ* generation of yeast to facilitate the fermentation process, and that spent yeast at the end of the ethanol process is a valuable nutritional component of DDGS. Indeed, the spent yeast of DDGS has been calculated to be approximately 6% of DDGS dry matter (Han and Liu 2010).

During fermentation, there is a significant net generation of protein from the yeast that is grown in the fermenter to facilitate fermentation. In addition to recovering corn protein, the MSC™ process also recovers a high proportion of spent yeast, in particular cell wall components. Whole yeast cells and yeast cell

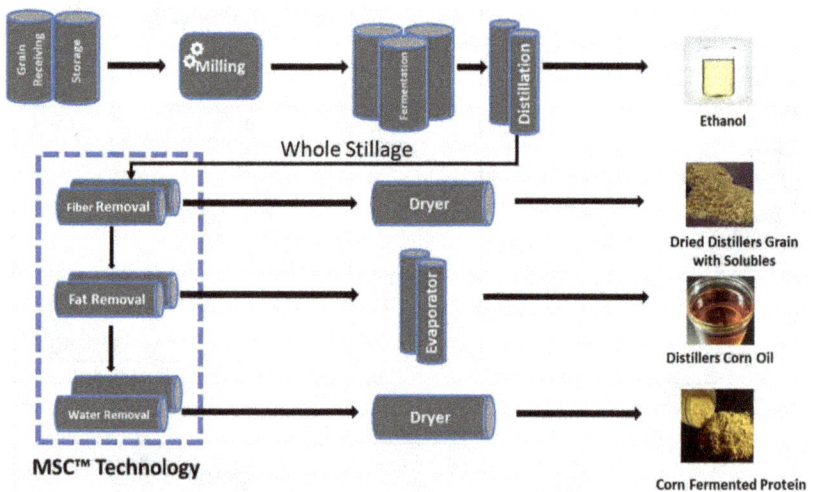

Figure 2 The maximized stillage co-products process.

components are recognized as valuable feed nutrient supplements in their own right (Shurson 2018). This produces an exceptionally high-protein CFP product (up to 60% crude protein) that has now been extensively and very successfully tested as a protein supplement for poultry.

A key challenge has been to measure the quantity of either yeast or yeast components in a dried material such as DDGS. The measurement of yeast cell wall mannose is accepted by the yeast industry as a proxy measure for yeast cell material. Yeast generated in fermentation accompanies stillage in the ethanol process and passes through the distillation columns where the yeast cells are lysed, releasing the intracellular components, including nucleotides. The components of yeast cell walls (mannan oligosaccharides and β-glucans) can be measured in dried CFP products. Using this proxy measure, the yeast cell wall material represents approximately 20–27% of the dry matter content of CFP. Approximately 9% of the protein in a 52% protein product is derived from yeast cell material. However, it is important to recognize that these spent yeast components are not equivalent to whole and viable yeast cells used as feed supplements, though the immuno-stimulatory activity of yeast cell components may be valuable functional components of CFP.

4 Challenges in producing corn-fermented protein

There are a number of challenges in producing a commercial CFP product. One potential safety issue is the use of antibiotics in the ethanol process. As a means of optimizing the efficiency of the yeast ethanol conversion step in the dry grind ethanol process, antibiotics have been used during fermentation to eliminate competitive bacterial activity. In the past, virginiamycin, erythromycin, and tylosin have been used routinely. Given that the stillage is exposed to high temperature during distillation, there is no residual antibiotic activity in the co-products, but antibiotic residues are detectable. Plants producing CFP are able to use a number of alternative products to antibiotics and, with attention to improved cleaning in place, they are able to operate without the use of antibiotics to control the fermentation process, allowing them to be compliant with No Antibiotics Ever (NAE) programs for food and feed production.

Another key aspect of product development is its environmental impact. CFP is produced in a highly competitive, sustainable manner with low GHG emission values compared to a range of different protein products (Fig. 3). Using CFP in ration formulation as a partial replacement of SBM has been reported to reduce the GHG value of feed formulation in diets for turkey poults (Burton et al., 2021).

Key to commercial sustainability is the ability to produce at scale. Up to 500 000 metric tons per annum of a specific formulated feed is a norm for commercial feed production. A batch of the formulated feed of up to 100

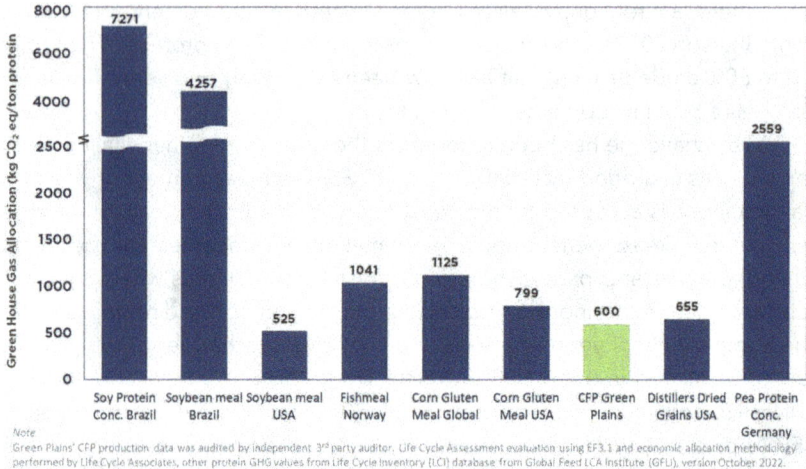

Figure 3 Greenhouse gas allocation values (kg CO_2 eq/ton protein) for a range of different protein products.

tons is made, then mixed in 10–12 ton mixer batches. Storage space for feed ingredients is limited. Generally, large-scale integrators are limited to three to four feed bins on-site, which limits the number of individual ingredients that can be stored. The availability of feed ingredients must merit bin space.

Key aspects of supply chain performance are resilience, redundancy, and reliability (the 3 R's). Changing a feed formulation, particularly for poultry, can be an expensive and time-consuming exercise. Resilience is key because feed producers will not rely on a single source of supply in case of a plant stoppage. It is essential to have one or more sites of manufacture of the product for there to be an alternative source of supply. For the same reason, there needs to be redundancy in volume production to account for increased demand. With resilience in the supply chain, there is an additional need for reliability (product consistency) across different suppliers. Producing a new novel feed protein ingredient is therefore not without significant challenge. CFP products are currently being produced from 17 different ethanol plants in the USA and Brazil with an annual production capacity close to one million tons. Production of the product in Europe is imminent.

Standard measures of product composition include amino acid digestibility using the precision-fed cecectomized rooster assay (PFR) (Parsons, 1985). The assay has been completed on nine separate occasions with samples of CFP obtained during the development of the product, from different ethanol plants as they came online over a period of approximately 6 years, in order to monitor the consistency of the product over time and from different plants. The ileal digestibilities of amino acids in CFP, as measured by the PFR technique in broilers, are presented in Table 1. Data from the full set of ten measurements of standardized digestibility is shown in Table 1 (Parsons et al., 2023), and the nutrient composition and TMEn of eight samples are presented in Table 2. The coefficient of variation of the *in vivo* mean standardized digestibility of the ten samples of CFP is 2.3%, demonstrating the excellent consistency in the nutritional *in vivo* value of CFP produced from different sites and over a period of approximately 6 years. Standardized protein digestibility values and TMEn values have been used to formulate CFP feed for broilers and turkey poults.

A number of different trials have been carried out to determine the efficacy of CFP when included as an alternative protein source in feed formulations for poultry.

5 Case study: corn-fermented protein as a feed for broilers

The goal of the study was to identify the effect of feeding CFP as a protein supplement for broilers in all phases of growth up to slaughter at 42 days of age. A total of 432 1-day old male Ross 308 birds were randomly allocated to four experimental groups:

- Control (fed on a commercial broiler concentrate for the whole of the trial);
- 5% CFP (control + 5% CFP for the whole of the trial);
- 10% CFP (control + 10% CFP for the whole of the trial); and
- 10% CFP 22+ (control + 5% CFP 22+ for the grower phase and control + for the finisher phase).

All diets were formulated to be equivalent in standardized digestible amino acid content and energy content, as well as nutrient requirements recommended for male Ross broilers. The diets were offered as crumbs in the starter phase and pellets in the grower phase. The birds had free access to feed and water and were reared in 0.8 m × 0.8 m pens, with 9 birds per pen, and pens were bedded with wood shavings. Growth performance data (live weight, live weight gain, feed intake, and feed conversion ratio) was recorded weekly (Table 3). At the age of 35 days, three birds were selected randomly from each pen, euthanized, and used to evaluate carcass yield (Table 3).

Ileal digesta was collected from four birds per pen (pooled) at D21 and D42. Ileal tissue for histology was collected from one bird/pen at D21. Villus height and crypt depth were measured, and the ratio of villus height to crypt depth was calculated. The nutritional treatments had no effect on histological measurements. Dry matter content of the excreta at day 21 and day 42 was not significantly affected by the dietary treatment. Bird growth and feed intake were

Table 1 Standardized ileal digestibility of amino acids in samples of corn-fermented protein

Columns CFP1–CFP8 are Univ Illinois samples 2017–2023; CFP9 columns are Univ Georgia.

Amino acid	CFP1 Digest vlue	CFP1 SEM[3]	CFP2 Digest value	CFP2 SEM[3]	CFP3 Digest value	CFP3 SEM[3]	CFP4 Digest value	CFP4 SEM[3]	CFP5 Digest value	CFP5 SEM[3]	CFP6 Digest value	CFP6 SEM[3]	CFP7 Digest value	CFP7 SEM[3]	CFP8 Digest value	CFP8 SEM[5]	CFP9 Digest value	CFP9 SEM[1]	CFP9	Mean
ALA	91.5	0.8	92.6	0.6	94.7	0.6	91.8	0.3	91.1	0.9	90.0	0.8	91.7	1.6	91.6	0.8	89.5	0.4	88.6	
ARG	93.3	1.0	96.6	0.8	98.1	0.8	93.2	0.5	93.6	0.9	91.8	0.9	93.6	2.3	92.3	0.6	91.3	0.5	91.5	
ASP	87.7	1.1	87.8	1.3	91.3	1.3	86.1	0.8	84.6	1.4	85.7	1.3	87.2	1.9	87.9	2.0	85.0	0.6	84.2	
CYS	86.6	2.2	86.3	2.2	91.8	2.2	85.7	1.1	86.5	1.9	78.6	2.1	81.2	3.7	78.9	2.6	79.8	1.5	79.5	
GLU	93.1	0.7	93.7	0.7	96.1	0.6	93.1	0.2	92.2	0.8	91.0	0.7	92.8	1.7	92.9	1.3	91.5	0.5	90.3	
GLY	–	–	–	–	–	–	–	–	–	–	–	–	–	–	–	–	68.9	–	52.5	
HIS	93.2	1.2	90.6	0.8	92.8	0.8	89.2	0.3	87.2	1.2	89.5	1.1	90.4	1.8	91.1	2.2	89.7	0.6	90.2	
ILE	93.8	0.8	91.9	0.7	93.2	0.7	90.5	0.2	89.9	1.0	87.3	0.9	89.3	2.0	89.3	1.0	87.6	0.4	86.3	
LEU	9*3	0.5	95.1	0.4	97.1	0.4	94.4	0.1	93.9	0.6	92.9	0.6	96.7	1.4	94.0	0.6	92.7	0.3	92.1	
LYS	83.4	1.5	83.2	1.1	90.5	1.1	81.0	0.8	80.0	2.3	85.5	1.6	88.9	2.4	89.2	2.4	80.5	0.9	90.7	
MET	91.5	0.4	93.3	0.5	96.9	0.5	92.3	0.1	92.0	0.9	92.8	0.6	93.9	1.7	93.9	0.7	91.6	0.4	90.5	
PHE	92.6	0.8	93.4	0.6	96.1	0.6	92.6	0.1	92.4	0.8	90.6	0.6	92.8	1.7	91.9	0.6	93.6	0.3	90.1	
PRO	92.4	0.9	92.4	1.1	95.4	1.1	92.0	0.2	90.9	1.0	88.5	0.7	91.1	1.9	89.7	1.0	93.8	0.7	90.2	
SER	88.5	1.7	89.6	1.9	93.0	1.9	89.1	1.0	88.2	1.3	86.5	1.1	89.2	2.8	87.7	1.1	36.6	0.9	87.9	
THR	86.1	1.5	87.0	1.6	89.9	1.6	85.7	1.2	83.5	1.6	85.5	1.1	87.1	2.6	87.2	2.3	86.2	1.0	86.1	
TRP	94.7	0.9	96.6	0.8	93.6	0.8	92.9	0.7	92.5	1.2	90.2	0.5	90.8	1.7	90.4	1.5	86.3	1.1	86.6	
TYR	93.3	0.9	96.6	0.6	96.9	0.6	93.1	0.2	92.8	1.0	90.8	0.8	92.1	1.8	91.9	1.1	91.9	0.55	91.6	
VAL	88.8	1.1	89.6	1.1	95.3	1.1	88.8	0.3	87.6	1.2	89.0	0.8	90.8	2.0	89.9	0.7	87.8	0.5	86.3	
Mean	**90.4**		**91.2**		**94.3**		**90.1**		**89.2**		**88.6**		**90.4**		**90.0**		**87.0**		**86.3**	**89.7**
ex GLY																	88.1		88.3	cv 2.3

Table 2 Composition and TMEn of corn-fermented protein (CFP) (DM basis)

	CFP1	CFP2	CFP3	CFP4	CFP5	CFP6	CFP7	CFP8
Crude protein (%)	54.8	57.7	50.7	57.8	57.7	54.3	56.5	56.3
Crude fat (%)	7.1	4.6	5.0	4.2	4.0	4.7	3.5	5.0
Ash (%)	5.9	4.7	2.3	4.3	4.1	4.6	3.2	3.1
Neutral detergent fiber (%)	27.7	29.1	39.7	35.6	32.3	24.1	36.7	37.8
Ca (%)	0.05	0.02	0.02	0.02	0.02	0.06	0.03	0.03
P (%)	1.29	0.78	0.47	0.76	0.77	1.04	0.81	0.82
Na (%)	0.09	0.08	0.13	0.08	0.08	0.09	0.04	0.04
Gross energy (kcal/g)	5.46	5.43	5.35	5.50	5.55	5.38	5.43	5.43
TMEn (kcal/g)	3.55	3.55	3.29	3.77	3.72	3.39	3.68	3.50
SEM of TMEn	0.08	0.10	0.04	0.04	0.17	0.05	0.21	0.11

not significantly affected by the dietary treatment. However, birds given the diet with 10% CFP had numerically higher feed intake and FCR compared to the controls. Furthermore, birds receiving 5% CFP and 10% CFP had significantly higher breast meat yield compared to the control.

A second trial (Burton et al., in press) was sponsored by a major European commercial poultry producer and was carried out to test the use of CFP in diets for commercial broilers (Burton, personal communication). The objectives of the study were to determine the effects on growth of broilers to differing levels of CFP inclusion in diets with and without an extruded canola protein product. A total of 512-day-old broilers (half male and half female) were randomly selected in a 2 × 2 factorial trial. Four diets were prepared:

- A control (commercial broiler formulation);
- A diet with 5% CFP added;
- A diet with 7.5%, CFP added; and
- A diet with 10.0% CFP added.

The CFP replaced SBM in iso-nitrogenous and iso-energetic diets. The experimental concentrate was in the form of crumb (D0-D10) and pellet (D11-D34). The birds had free access to feed and water and were reared in 0.8 m × 0.8 m pens with wood shavings as bedding. Growth performance (live weight, live weight gain, feed intake, and feed conversion ratio) was recorded every 7 days during the trial. The litter score was recorded at D31, and litter samples were collected for dry matter determination (Table 4). At D21, two birds/pen were euthanized by cervical dislocation, and ileal digesta content was collected. Both ceca from one bird per pen were collected, snap-frozen immediately, and then stored at −80°C until analysis of volatile fatty acids (VFAs) was performed.

Table 3 Effect of corn fermented protein supplementation on growth performance and carcass yields of broilers: Cumulative performance

		Treatments				
Bird weight gain (g)	Control	5% CFP	10% CFP	10% CFP 22+	SEM	P value
D14	499	503	518	501	5.73	0.088
D21	1043[b]	1050[b]	1081[a]	1067[ab]	7.12	0.003
D28	1785[b]	1841[ab]	1899[a]	1832[ab]	23	0.014
D35	2539	2635	2652	2606	30.7	0.064
D42	3315	3394	3295	3304	32.7	0.144
Feed intake (g)						
D14	604	625	624	632	7.38	0.059
D21	1333[b]	1354[ab]	1365[ab]	1377[a]	9.16	0.013
D28	2436	2486	2527	2518	30.9	0.172
D35	3657[b]	3764[ab]	3839[a]	3792[ab]	44.6	0.044
D42	4878[b]	5042[ab]	5151[a]	5066[ab]	61.8	0.028
Feed conversion ratio						
D14	1.21[bc]	1.24[ab]	1.20[c]	1.26[a]	0.01	0.001
D21	1.28	1.29	1.26	1.29	0.007	0.055
D28	1.37[a]	1.35ab	1.33b	1.38a	0.008	0.002
D35	1.44	1.43	1.45	1.46	0.012	0.408
D42	1.47[b]	1.49[b]	1.57[a]	1.53ab	0.02	0.005
Carcass yield (g)						
Breast	742[c]	826[a]	797[ab]	771[bc]	14.1	0.001
Drumstick	298	300	296	286	4.63	0.145
Thigh	365	362	356	345	5.93	0.114

Means within a row with similar superscript are not significantly different at $P \leq 0.05$. $n = 12$
SEM, standard error of the mean

On day 34, two birds/pen were euthanized by cervical dislocation to record carcass yield. Litter score was recorded at D31, and litter samples were collected for dry matter determination. Caecal contents were collected for VFA analysis at D21 and D42 from one bird per pen and snap-frozen immediately on dry ice before storing at −80°C. Quantification of short-chain fatty acids in mammalian feces in aqueous solution content was carried out using gas chromatography–mass spectrometry (GC-MS) with a Shimadzu 2010ultra GC-MS single quadrupole mass spectrometer equipped with a Shimadzu GC2010 and AOC20i+s autosampler (Shimadzu Corp., Kyoto, Japan). The results indicate that the presence of CFP in the diets had significant effects on the proportions of individual short-chain volatile fatty acids in the cecal contents of the birds, and there was a tendency for total VFA concentration in cecal content to be increased (Table 5). The presence of CFP in the diet was influencing and increasing cecal fermentation.

The use of CFP was further evaluated in boilers in a second commercial trial using 4000 Hubbard M99 × Cobb 500 chicks supplied from a common breeder flock. The trial was designed to evaluate the live performance of broilers fed a standard small broiler control corn–soybean meal diet with diets containing CFP at inclusion levels: 3%; 6%; 9% used to replace SBM.

Table 4 Growth performance, meat yield and litter dry matter of male and female broilers

	Control	Extrupro 10%	CFP 5%	CFP 7.5%	CFP 10%	7.5% CFP 7.5% + Extrupro 10%
Female birds						
D0–D34 bird weight gain (g)	2124	2234	2196	2187	2188	2149
D0–D34 feed intake (g)	3084	3117	3105	3116	3171	3189
D0–D34 FCR	1.37	1.35	1.33	1.33	1.38	1.39
Meat yield (g)	844b	890ab	888ab	917a	896ab	902ab
Litter dry matter (%)	72.3b	73.8ab	75.8ab	73.9ab	76ab	76.4a
Male birds						
D0–D34 bird weight gain (g)	2485	2574	2511	2513	2495	2464
D0–D34 feed intake (g)	3507	3551	3546	3530	3606	3467
D0–D34 FCR	1.33	1.3	1.33	1.31	1.35	1.31
Meat yield (g)	1063ab	1055ab	1019b	1065ab	1093a	1051ab
Litter dry matter (%)	69c	70.5bc	74.7ab	75.8a	71.9abc	74.6ab

Means within a row with similar superscript are not significantly different at $P \leq 0.05$. $n = 32$

All diets were balanced for protein, standardized amino acid content, and TMEn content. With n = 8 replicates per treatment and 120 chicks per pen, the treatments were randomized within each block. All feeds were pelleted, and birds received starter diets from 1 to 14 days, finisher diets from 15 to 25 days, and withdrawal diets from 26 to 33 days of age. All birds were weighed at 33 days of age, and average weights and feed conversion ratios were calculated for each pen. The average final weight (kg) of the control 0, 3% CFP, 6% CFP, and 9% CFP were 2.22, 2.14, 2.15, and 2.12 respectively, and FCR 1.49, 1.50, 1.49, and 1.50 respectively, with mortality in the groups 0.5, 1.4, 1.0, and 1.5. There were no significant differences in growth, feed conversion, or mortality between the four treatments. The results indicated that CFP can be used as a partial replacement for SBM in the diet of broilers.

6 Case study: corn-fermented protein as a feed for turkey poults

Turkey poults have a reputation for having a highly sensitive digestive tract, and providing an appropriate starter diet is key to optimizing performance. A comparison was made between the use of SBM and partial replacement with CFP in starter diets for turkey poults (Scholey et al., 2023). CFP was used to replace approximately 6% and 12% of the SBM in the diets, with 4% and 8% of CFP, respectively. To formulate the diets, standardized ileal digestibility and TMEn were based on evaluations made in cecetomized cockerels (Table 1). All diets were formulated to commercial specifications, with pellet integrity between 98% and 99.3% for all experimental diets. The pellet quality of the experimental diet was not influenced by the inclusion of CFP, and CFP replaced approximately 6% and 12% of the SBM in the diets (4% and 8% CFP inclusion, respectively). Litter quality and foot pad score were assessed as simple indicators of gut health. The bedding remained friable and dry throughout the study, and no extra bedding was added to pens on any treatments, suggesting that there was no adverse effect of dietary treatment on gut health. The growth performance of poults from 0 to 42 days of age across all treatments aligned well with industry performance targets (Aviagen, 2014), indicating that the diets conformed well to industry standards (Table 6).

Although the five diets were formulated to supply equivalent quantities of standardized digestible amino acids and AMEn, poults fed the CFP8% diet were significantly heavier (plus ~8.2%) and gained more weight (plus ~8.4%) than those fed the control diet based on SBM (Table 6). However, there was no significant difference in feed intake. Based on the nutrient availability measured in the poults at the termination of the trial, poults offered CFP8% had significantly higher AME, AMEn, and N retention over the control poults by ~26%, 27%, and 13%, respectively. Moreover, the digestibility of VAL and ILE of CFP8% poults

Table 5 Effect of including CFP in the diets on hind gut volatile fatty acid content a+B6:L22nd composition

	Formic acid	Acetic acid	Propanoic acid	Isobutyric acid	Butyric acid	Iso valeric acid	Valeric acid	Iso. hexanoic acid	Hexanoic acid	Heptanoic acid
Female										
CONTROL	129[a]	5393[b]	374[b]	46[b]	1912[b]	25.7[b]	84.6[b]	15.2[a]	19.4[b]	19.6[a]
CFP 10%	149[a]	7099[a]	502[a]	75.4[a]	2785[a]	38.1[a]	136.7[a]	16.7[a]	23.8[a]	21.5[a]
Male										
CONTROL	149[a]	4762[a]	327[b]	45.5[b]	1576[a]	25.6[a]	85[b]	17.9[a]	21[a]	23.4[a]
CFP 10%	136[a]	5425[a]	459[a]	63[a]	1636[a]	34.7[a]	108.9[a]	16.5[a]	19.2[a]	20.8[a]
SEM	9.42	509	36.8	4.24	220	3.42	6.21	0.883	0.991	1.14
P value										
Diet	0.798	0.027	0.001	<0.001	0.044	0.003	<0.001	0.884	0.249	0.76
Sex	0.62	0.032	0.228	0.172	0.002	0.634	0.042	0.179	0.169	0.187
Diet × Sex	0.092	0.314	0.946	0.17	0.076	0.636	0.032	0.117	0.004	0.061

Means within a row with similar superscript are not significantly different at $P \leq 0.05$; n = 32
CFP10% = 10% of the total diet was corn fermented protein

were higher than the control poults (by ~11% and ~7%, respectively). Since there was no effect on feed intake, it is reasonable to conclude, therefore, that partial replacement of SBM by an apparently equivalent supply of protein and energy from CFP actually resulted in an increased supply of protein and energy to the poults. Thus, the supply of protein and energy of CFP8% was higher compared to the controls, which might explain the improvement in poult weight gain in CFP8% poults.

Both the CFP 4% and CFP8% diets had significantly higher AME compared to the control. In addition, CFP8% had significantly higher N retention compared to the control (Table 7).

7 Conclusion

There are currently 17 plants producing CFP in the USA, two in South America, and one in construction in Europe, with a combined production capacity of well over one million tons of CFP per annum, providing resilience and redundancy in the supply chain. Although the dry grind ethanol process is a global source of medium-quality protein, the opportunity to turn corn into a multi-purpose energy and protein crop is a major breakthrough. The trials carried out to evaluate the use of CFP as an alternative protein supplement in diets for broilers and turkey poults demonstrate that CFP can be used in the formulation as an alternative protein.

In order to complete these trials, standardized digestibility and true metabolizable energy were measured in the target species. In all the growth experiments, the maximum inclusion rate of CFP was restricted to not exceed 10% of dietary inclusion. Since these experiments were exploratory experiments to test the use of the product in diets for poultry and CFP contains approximately 20–27% yeast cell wall material, this was a precautionary measure to not exceed the yeast cell wall dietary inclusion level of 5%. CFP can be used as a dietary replacement for SBM, and with appropriate balancing of essential amino acids, it will achieve performance at least equal to the performance of the control diets. However, there is evidence to suggest that either the content of spent yeast material or the combination of spent yeast with a fermented carbohydrate fraction may be beneficial in terms of gut health.

The trial with turkey poults is particularly interesting. Overall, the growth performance of poults across all treatments aligned well with industry standards (Aviagen, 2014), indicating that the diets conformed well to industry standards. However, in terms of improved nitrogen end energy retention, there is evidence to suggest that there may be additional benefits to the partial replacement of SBM with CFP.

CFP contains a functional component of approximately 24% (on a DM basis) spent yeast content. Several nutraceutical components of yeast are

Table 6 Effect of corn-fermented protein on body weight, feed intake, feed conversion ratio and litter dry matter of turkey poults

	Control	CFP4%	CFP8%	Econ + CFP4%*	Premium	SEM	P value
D0 bird weight	66.2	66.2	65.6	65.5	65.3	1.05	0.962
D42 bird weight (g)	2328b	2423ab	2518a	2357b	2430ab	52.1	0.011
D0–D42 bird weight gain (g)	2262b	2357ab	2452a	2292ab	2365b	60.2	0.009
D0–D42 feed intake (g)	3741	3850	3743	3756	3842	77.3	0.363
D0–D42 feed conversion ratio	1.66	1.64	1.61	1.64	1.63	0.028	0.797
D42 litter dry matter	74	73.7	73.9	75.5	75.1	1.15	0.752
Carbon footprint (kg CO2 eq/kg bird weight gain)	4.31a	4.18a	3.99b	3.59c	2.85d	0.066	<0.001

*CFP added in economy diet at 4%

Means within a row with similar superscript are not significantly different at $P \leq 0.05$

SEM, standard error of the mean

Control = High protein with soybean meal as primary protein source

CFP4% = 4% of the total diet was corn fermented protein used in place of high protein soybean meal

CFP8% = 8% of the total diet was corn fermented protein used in place of high protein soybean meal

Econ+CFP = 4% of the total diet was corn fermented protein with lower protein, higher fiber soybean meal, as primary protein source

Premium = 5% of the total diet was a soy protein isolate in place of high protein soybean meal

N = 10 with 5 birds per pen

Table 7 Effect of corn fermented protein on apparent metabolizable energy, apparent metabolizable energy corrected for nitrogen and nitrogen retention of turkey poults

	AME (MJ/kg)	N retention (g/ kg of the diet)	AMEn (MJ/kg)
Control	10.1b	23.3b	9.3b
CFP4%	12.4a	25.3ab	11.6a
CFP8%	12.7a	26.3a	11.8a
Econ+CFP4%	9.7b	20.0c	9.0b
Premium	12.1a	23.5b	11.3a
SEM	0.54	0.86	0.52
P value	<0.001	<0.001	<0.001

*CFP added at 4% in economy diet
[a-c] Means within the same column with no common superscript differ significantly ($P \leq 0.05$)
SEM, standard error of the mean
Control = high protein with soybean meal as primary protein source
CFP4% = 4% of the total diet was corn fermented protein replacing soybean meal
CFP8% = 8% of the total diet was corn fermented protein replacing soybean meal
Econ+CFP4% = 4% of the total diet was corn fermented protein with lower protein, higher fiber soybean meal, as primary protein source
Premium = 5% of the total diet was a soy protein isolate replacing soybean meal protein
AME = Poultry apparent metabolizable energy
AMEn = apparent metabolizable energy corrected for nitrogen

reported to have beneficial effects in poultry and may indirectly improve growth performance. β-1,3-glucans, the major component of yeast cell walls, have prebiotic effects due to their ability to bind toxins and pathogens (Vetvicka et al., 2014). Yeasts contain mannan oligosaccharides which have an overall positive effect on animal performance (Spring et al., 2015) via improving intestinal architecture, physical gut tissue turnover, changes in microbiota, or reduction in immune stimulation (Baurhoo et al., 2007). The effects on cecal VFA concentrations reported in the broiler trial are evidence that the fermented fiber component of CFP has the potential to increase hindgut fermentation although to date, there is no evidence to indicate an improvement in the histology of the gut.

Identifying new sources of high-quality vegetable protein available for use in animal feed is a priority. High-quality vegetable proteins are extensively used for food and can therefore demand a price premium. Identifying a source of high-concentration vegetable protein, such as CFP, which is not available for use in food since it is derived from a non-food industrial process, is an opportunity for the feed industry to source an alternative cost competitive product. While SBM is recognized as an excellent protein source, residual trypsin inhibitor, lipoxygenases, and antigenic proteins retain a degree of activity post-processing, which slightly but significantly reduces dietary protein digestion (Clarke and Wiseman, 2005). Thus, some of the performance improvements in the current study could be related to the reduction in anti-nutritional effects associated with partial replacement of SBM with CFP. However, it is important to acknowledge the potential limitations of working with protein fractionation co-products derived from the ethanol industry. The ethanol industry gains significant financial benefits from pursuing oil recovery. Fractionation of whole stillage significantly increases the efficacy of oil recovery, which reduces the energy content of the resultant co-product. It is essential, therefore, to refer to the declared energy level in the co-products derived from different processes. In addition, corn protein products will have significantly higher levels of isoleucine and leucine compared to an equivalent level of protein from SBM. Indeed, the ratio of leucine plus isoleucine to lysine is approximately 6× greater in corn compared to an equivalent soybean protein-based product. Acknowledging potential amino acid imbalances, which can be rectified with amino acid supplementation, and differences in energy content is essential to achieve maximum performance from corn-based ethanol co-products.

The current program covering the use of CFP in poultry is part of an extensive program on the use of protein in diets for swine, aquaculture, companion animals, and ruminants, in addition to poultry. The results reflect the beneficial effects reported in poultry. The development of CFP is an excellent example of how the introduction of tested mechanical processing technology with existing commodity feed materials can significantly improve the value of low-quality feed material and increase the flexibility and value of feed materials used by animal nutritionists in feed formulation. There have been few if any, new and significant developments in the vegetable protein space for animal feed in the past 20 years. CFP is a major development in terms of a new commercially viable high-protein feed for all classes of livestock and companion animals.

8 References

Aviagen 2014. *Ross 308 Broiler: Nutrition Specifications* [WWW Document]. http://en.aviagen.com/assets/Tech_Center/Ross_Broiler/Ross308BroilerNutritionSpecs2014-EN.pdf (accessed December 12, 2016).

Baurhoo, B., Phillip, L. and Ruiz-Feria, C. A. 2007. Effects of purified lignin and mannan oligosaccharides on intestinal integrity and microbial populations in the ceca and litter of broiler chickens. Poultry Science 86(6). 1070–1078. https://doi.org/10.1093/ps/86.6.1070.

Xu, B., Li, Z., Wang, C., Fu, J., Zhang, Y., Wang, Y. and Lu, Z. 2020. Effects of fermented feed supplementation on pig growth performance: a meta-analysis. Animal Feed Science and Technology 259. 114315. https://doi.org/10.1016/j.anifeedsci.2019.114315.

Burton, E., Scholey, D., Alkhtib, A. and Williams, P. 2021. Use of an ethanol bio-refinery product as a soy bean alternative in diets for fast-growing meat production species: a circular economy approach. *Sustainability* 13(19). 11019.

Burton, E., Scholey, D., Alkhtib, A. and Williams, P. Alternative in diets for fast-growing meat production species. *Sustainability* 13(19). 11019.

Clarke, E. and Wiseman, J. 2005. Effects of variability on trypsin inhibitor content of soy bean meals on true and apparent ileal digestibility of amino acids and pancreas size in broiler chicks. *Animal Feed Science and Technology* 121(1-2). 125-138. https://doi.org/10.1016/j.anifeedsci.2005.02.012.

Han, J. and Liu, K. 2010. Changes in composition and amino acid profile during dry grind ethanol processing from corn and estimation of yeast contribution towards DDGS proteins. *Journal of Agricultural and Food Chemistry* 58(6). 3430-3437.

Parsons, C. M. 1985. Influence of caecectomy on digestibility of amino acids by roosters fed distillers dried grains with solubles. *Journal of Agricultural Science* 104(2). 469-472.

Parsons, B. W., Utterback, P. L., Parsons, C. M. and Emmert, J. L. 2023. Standarized amino acid digestibility and true metabolizable energy for several increased protein ethanol co-products using back-end fractionation systems. *Poultry Science* 102(2). 102329.

Mottet, A., de Haan, C., Falcuucci, A., Tempio, G., Opio, C. and Gerber, P. 2017. Livestock: on our plates or eating at our table? A new analysis of the feed/food debate. *Global Food Security* 14. 1-8. https://doi.org/10.1016/j.gfs.2017.01.001.

Scholey, D., Alkhtib, A., Wiliams, P. and Burton, E. 2023. Corn fermented protein, an alternative protein to soybean meal to support growth in young turkey poults. *Journal of Applied Animal Nutrition.* https://doi.org/10.3920/JAAN2023.0002.

Shurson, G. C. 2018. Yeast and yeast derivatives in feed additives and ingredients: sources, characteristics, animal responses and quantification methods. *Animal Feed Science and Technology* 235. 60-76. https://doi.org/10.1016/j.anifeedsci.2017.11.010.

Spring, P., Wenk, C., Connolly, A. and Kiers, A. 2015. A review of 733 published trials on Bio-Moss, a mannan oligosaccharide and Actigen, a second generation mannose rich fraction, on farm and companion animals. *Journal of Applied Animal Nutrition* 3. 1-11.

Vetvicka, V. and Oliveira, C. 2014. β(1-3)(1-6)-D-glucans modulate immune status in pigs: potential importance for efficiency of commercial farming. *Annals of Translational Medicine* 2(2). 16. https://doi.org/10.3978/j.issn.2305-5839.2014.01.04.

Williams, P. E. V. 2023. High protein corn products for swine derived from corn ethanol production. In press.

www.ingramcontent.com/pod-product-compliance
Lightning Source LLC
Chambersburg PA
CBHW050536270326
41926CB00015B/3248